HISTORICAL AND PHILOSOPHICAL DIMENSIONS OF LOGIC, METHODOLOGY AND PHILOSOPHY OF SCIENCE

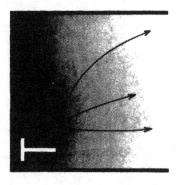

THE UNIVERSITY OF WESTERN ONTARIO
SERIES IN PHILOSOPHY OF SCIENCE

A SERIES OF BOOKS

ON PHILOSOPHY OF SCIENCE, METHODOLOGY,

AND EPISTEMOLOGY

PUBLISHED IN CONNECTION WITH

THE UNIVERSITY OF WESTERN ONTARIO

PHILOSOPHY OF SCIENCE PROGRAMME

VOLUME 12

HISTORICAL AND PHILOSOPHICAL DIMENSIONS OF LOGIC, METHODOLOGY AND PHILOSOPHY OF SCIENCE

PART FOUR OF THE PROCEEDINGS
OF THE FIFTH INTERNATIONAL CONGRESS OF
LOGIC, METHODOLOGY AND PHILOSOPHY OF SCIENCE,
LONDON, ONTARIO, CANADA–1975

Edited by

ROBERT E. BUTTS

The University of Western Ontario

and

JAAKKO HINTIKKA

The Academy of Finland and Stanford University

D. REIDEL PUBLISHING COMPANY
DORDRECHT-HOLLAND/BOSTON-U.S.A.

Library of Congress Cataloging in Publication Data

International Congress of Logic, Methodology, and Philosophy of Science, 5th, Univer-
    sity of Western Ontario, 1975.
    Historical and philosophical dimensions of logic, methodology, and philosophy of
    science.

    (Proceedings of the Fifth International Congress of Logic, Methodology, and
Philosophy of Science, London, Ontario, Canada, 1975; pt. 4) (University of Western
Ontario series in philosophy of science; v. 12)
    Bibliography: p.
    Includes index.
    1. Science—Philosophy—History—Congresses. 2. Science—Methodology—
History—Congresses. 3. Logic, Symbolic and mathematical—History—Con-
gresses. I. Butts, Robert E. II. Hintikka, Kaarlo Jaakko Juhani, 1929-
   III. Title. IV. Series: University of Western Ontario, The University of Western
Ontario series in philosophy of science; v. 12.
Q174.I58 1975a pt. 4   [Q174.8] 501s [501]   77-22433
ISBN 90-277-0831-2

Published by D. Reidel Publishing Company,
P.O. Box 17, Dordrecht, Holland

Sold and distributed in the U.S.A., Canada, and Mexico
by D. Reidel Publishing Company, Inc.,
Lincoln Building, 160 Old Derby Street, Hingham,
Mass. 02043, U.S.A.

Printed in The Netherlands

TABLE OF CONTENTS

PREFACE

The Fifth International Congress of Logic, Methodology and Philosophy of Science was held at the University of Western Ontario, London, Canada, 27 August to 2 September 1975. The Congress was held under the auspices of the International Union of History and Philosophy of Science, Division of Logic, Methodology and Philosophy of Science, and was sponsored by the National Research Council of Canada and the University of Western Ontario. As those associated closely with the work of the Division over the years know well, the work undertaken by its members varies greatly and spans a number of fields not always obviously related. In addition, the volume of work done by first rate scholars and scientists in the various fields of the Division has risen enormously. For these and related reasons it seemed to the editors chosen by the Divisional officers that the usual format of publishing the proceedings of the Congress be abandoned in favour of a somewhat more flexible, and hopefully acceptable, method of presentation.

Accordingly, the work of the invited participants to the Congress has been divided into four volumes appearing in the University of Western Ontario Series in Philosophy of Science. The volumes are entitled, *Logic, Foundations of Mathematics and Computability Theory, Foundational Problems in the Special Sciences, Basic Problems in Methodology and Linguistics,* and *Historical and Philosophical Dimensions of Logic, Methodology and Philosophy of Science.* By means of minor rearrangement of papers in and out of the sections in which they were originally presented the editors hope to have achieved four relatively self-contained volumes.

Most of the papers appearing in this volume consist of those submitted by invited participants working in the history of logic, methodology and philosophy of science. Exceptions are the papers by Earman and Strauss (originally read in the section devoted to foundations of physics), Agazzi and Supek (originally read in the section on general methodology), and the symposium papers on theory change (also originally included in the section on general methodology). In each case, inclusion in this work was motivated by the attempt to produce a group of papers related in subject-matter and in general field of interest. Contributed papers in the areas of scholarship represented in this volume appeared in the volume of photo-offset preprints distributed at the Congress. The full programme of the Congress is printed as an appendix to this book.

The work of the members of the Division was richly supported by the National Research Council of Canada and the University of Western Ontario. We here thank these two important Canadian institutions. We also thank the Secretary of State Department of the Government of Canada, Canadian Pacific Air, the Bank of Montreal, the *London Free Press*, and I.B.M. Canada for their generous support. Appended to this preface is a list of officers and those responsible for planning the programme and organizing the Congress.

THE EDITORS

February 1977

OFFICERS OF THE DIVISION

A. J. Mostowski	(Poland)	President
Jaakko Hintikka	(Finland)	Vice President
Sir A. J. Ayer	(U.K.)	Vice President
N. Rescher	(U.S.A.)	Secretary
J. F. Staal	(U.S.A.)	Treasurer
S. Körner	(U.K.)	Past President

PROGRAMME COMMITTEE

Jaakko Hintikka (Finland), Chairman
R. E. Butts (Canada)
Brian Ellis (Australia)
Solomon Feferman (U.S.A.)
Adolf Grünbaum (U.S.A.)
M. V. Popovich (U.S.S.R.)
Michael Rabin (Israel)

Evandro Agazzi (Italy)
Bruno de Finetti (Italy)
Wilhelm Essler (B.R.D.)
Dagfinn Føllesdal (Norway)
Rom Harré (U.K.)
Marian Przełecki (Poland)
Dana Scott (U.K.)

CHAIRMEN OF SECTIONAL COMMITTEES

Y. L. Ershov (U.S.S.R.)	Section I:	Mathematical Logic
Donald A. Martin (U.S.A.)	Section II:	Foundations of Mathematical Theories
Helena Rasiowa (Poland)	Section III:	Computability Theory
Dagfinn Føllesdal (Norway)	Section IV:	Philosophy of Logic and Mathematics
Marian Przełecki (Poland)	Section V:	General Methodology of Science
J.-E. Fenstad (Norway)	Section VI:	Foundations of Probability and Induction
C. A. Hooker (Canada)	Section VII:	Foundations of Physical Sciences

Lars Walløe (Norway)	Section VIII:	Foundations of Biology
Brian Farrell (U.K.)	Section IX:	Foundations of Psychology
J. J. Leach (Canada)	Section X:	Foundations of Social Sciences
Barbara Hall Partee (U.S.A.)	Section XI:	Foundations of Linguistics
R. E. Butts (Canada)	Section XII:	History of Logic, Methodology and Philosophy of Science

LOCAL ORGANIZING COMMITTEE

R. E. Butts (Philosophy, the University of Western Ontario), Chairman

For the University of Western Ontario:

Maxine Abrams (Administrative Assistant)
R. N. Shervill (Executive Assistant to the President)
G. S. Rose (Assistant Dean of Arts)
R. W. Binkley (Philosophy)
J. J. Leach (Philosophy)
C. A. Hooker (Philosophy)
J. M. Nicholas (Philosophy)
G. A. Pearce (Philosophy)
W. R. Wightman (Geography)
J. D. Talman (Applied Mathematics)
J. M. McArthur (Conference Co-ordinator)

For the City of London:
Betty Hales (Conference Co-ordinator)

For the National Research Council of Canada:
R. Dolan (Executive Secretary)

I

HISTORY OF LOGIC, METHODOLOGY AND PHILOSOPHY OF SCIENCE

LARRY LAUDAN

THE SOURCES OF MODERN METHODOLOGY

I. INTRODUCTION

Scarcely two decades ago, those scholars interested in the history of the philosophy of science had to spend the bulk of their time justifying the existence of this field as a proper and legitimate branch of philosophy. Although prepared to concede that metaphysics, ethics and epistemology all had distinguished temporal careers, laden with present-day significance, most philosophers of science were convinced that philosophy of science really began during their own lifetimes, probably in Vienna; if it had any prior ancestry at all, it was generally traced no further back than Duhem, Mach, and Poincaré, along with occasional footnotes to Hume and Aristotle. The small coterie of scholars who were convinced that the story was a bit more complicated generally suppressed their internal disagreements and closed ranks – at least in public – in order to persuade their philosophical colleagues that the history of methodology was a florishing, exciting and relevant area of inquiry which deserved serious study and attention.

The litany of arguments invented to establish the importance of historical research in this area is, by now, a familiar refrain: we argued that philosophy of science is a subject with an active and interesting past; we pointed out that many problems treated in the historical literature (e.g. in the Whewell-Mill debates, or in the controversy between Mach and Boltzmann about theoretical entities) are pertinent and germane to contemporary discussions; in our less glowing moments, we even fell back on the old cliché about understanding the past in order not to repeat its mistakes.

On the whole, that battle has now been won. No longer do we see the outrageous historical nonsense that used to be commonplace: Hempel and Popper are no longer credited with discovering the

Butts and Hintikka (eds.), Historical and Philosophical Dimensions of Logic, Methodology and Philosophy of Science, 3-19.
Copyright © 1977 by D. Reidel Publishing Company, Dordrecht-Holland. All Rights Reserved.

covering-law model of explanation; Reichenbach and Carnap are no longer regarded as the first thinkers to apply mathematical probability to inductive inference; it is no longer taken for granted that comparative theory evaluation is a problem invented in the 20th century.

The signs of victory go much deeper, however, than an emerging historical punctiliousness would suggest. Both in terms of quantity and quality, studies devoted to the history of methodology have increased perceptibly in the last 15 years. Major philosophers of science within the field – men and women whose predecessors had little time for historical studies – are turning increasingly to look seriously at the evolution of methodological ideas.

Given these trends, given that we have finally won a sympathetic ear, it is time for those of us who are serious about the history of methodology to admit that all is not well. We should own up to the fact that the scholarly standards, as well as the historiographical assumptions, which have directed much of the research in this area are ill-considered and badly-conceived. What understanding we do have of the past is sketchy, disconnected and episodic. Before we move ahead with detailed and specific projects, it is time we examined publicly some of those well-concealed skeletons which have privately troubled many of us for sometime.

The quickest way to bring some of these skeletons out in the open is to ask, at least rhetorically, a number of elementary historical questions: Why was eliminative induction abandoned in the 19th century? When and why did the doctrine of thought experiments become explicit? What was the historical connection between Comtean positivism and logical positivism? Why did instrumentalism become fashionable in the late 19th century? How did the conceptions of scientific law or hypothesis evolve? When, and for what reasons, did philosophers of science first recognize that the positive instances of a theory were not all of equal confirmatory value? When and where did the verifiability theory of meaning originate?

For all their differences, these questions have two crucial features in common: (a) they are questions to which any adequate history of scientific epistemology should presumably provide answers; (b) *none* of them have yet been answered to anyone's satisfaction. These and a host of other similar queries about the genesis and temporal careers of the core concepts of modern methodology have had little light shed on them by the scholarly literature. There is no single methodological concept (from induction to explanation, from reduction to experiment) and no single methodological school (from instrumentalism to confirmationism, from positivism to pragmatism) whose history has yet been satisfactorily unravelled.

Thirty years ago, one might have put this down to a paucity of studies of the historical literature and the limited scholarly activity devoted to the field; but that excuse is beginning to wear thin, considering that more than a thousand books and articles have appeared in this field since 1950.[1] I am inclined to think that the failure of recent scholarship to produce many interesting answers to questions such as those listed above betrays some serious flaws in the presuppositions which most of us, myself included, have brought to the writing of history.

It is those flaws, and a prescription for their cure, which form the topic of this paper.

II. A GENERAL STATEMENT OF THE PROBLEM

In its most general form, the problem can be succinctly put: we have brought to the writing of the history of methodology certain preconceptions which jointly render it almost impossible to understand the evolution of this subject. These preconceptions concern both the nature of philosophy and the aims of history. Until those conceptions are altered, I venture to claim, the history of methodology will not deserve the serious attention which it would otherwise warrant.

I want to talk about this family of related philosophico-historical conceptions, using some specific examples to indicate the problem and to underscore its acuteness.

The most pernicious assumptions now afflicting the historiography of scientific epistemology are a set of persistent, but rarely explicit, theses about the character of the philosophy of science. Ironically, we have yet to learn the very lesson that we, as philosophers of science, have been preaching for more than a decade to our colleagues in the history of science. For just as it is true that one's conception of what science is affects how one conceives the history of science, so, too, does one's conception of the scope and aims of the philosophy of science condition one's approach to the history of methodology. Meta-level assumptions about what constitutes philosophy of science (based on hasty generalizations from what philosophy of science *now* is) can and have significantly affected the historical problems we explore and have drastically delimited the range of explanatory options open to us as historians.

This phenomenon of 'backwards projection' is, of course, endemic to intellectual history generally and is thus scarcely unique to the history of the philosophy of science. But what makes the problem especially acute for us is the fairly radical divergence between contemporary philosophy of science and its ancestors. It is just those divergences, which I shall discuss at length shortly, which make it very misleading to read the past through contemporary spectacles. There are, I believe, two principal problems here:

(1)   The first, and most chronic, assumption is that the philosophy of science is, above all, a branch of philosophy: that its problems, its practitioners, and its doctrines are drawn primarily from the area of general philosophy. The philosopher of science is perceived as an applied epistemologist, whose primary intellectual concerns and affiliations are traceable to the general philosophical climate in which he works.

The borderline between the philosopher and the philosopher of science becomes on this view very hazy; and it is presupposed that the most important and original philosophers are likewise profound philosophers of science and *vice versa*.

(2)   It is also assumed that the philosopher of science is a predominately normative theorist, whose doctrines are neither beholden to, nor determined by, the vagaries of contemporary science. The suggestion that the philosopher of science might engage in *ad hoc* special pleading for one current scientific theory at the expense of another is anathema to the carefully-cultivated image of the philosopher who manages to transend the cut and thrust of scientific partisanship.

However appropriate these views might be when applied to the present, however desirable they might be in principle, their transposition into the historiography of scientific epistemology has had disastrous consequences, for they make difficult the recognition that, so far as the past is concerned, philosophy of science has had a very different role and a very different set of disciplinary orientations than it now exhibits.

These two assumptions about the relation of philosophy of science to science constitute in one guise or another, the unstated, but pervasive model which historians of methodology have tended to graft onto the past, producing what I shall call the *'purist' historiography of method*. Histories written in this 'purist' mold work from two central assumptions:

(a) that one generally limits one's historical inquiries to those thinkers who already loom large in the history of general philosophy;

(b) that one seeks to explain the major developments in the history of methodology entirely with respect to the *philosophical* merits and deficiencies of certain ideas, rather than looking to the non-philosophical roots of many methodological ideas. Examples of historical studies premised on these two doctrines fill the journals and the library shelves. If I hesitate to cite any one by name, it is only because that would be to single out for criticism a work which shares an almost universal flaw.

What is so dreadful about the purist historiography? For starters, its palpable falsity! As I shall try to exhibit at length in what follows, philosophy of science has traditionally stood in a very different relation both to science and to philosophy from the one now familiar to us. This contrast makes nonsense of the effort to explain the historical evolution of methodology in the same terms in which we would discuss its very recent past. I shall claim that the simplistic projection of certain features of contemporary philosophy of science into the past has rendered it impossible to understand or explain many of the signal developments in the development of scientific epistemology, developments which should be the central foci of the historian. I shall claim, further, that, until we recognize that earlier philosophers of science had explanatory ambitions and motivations which were, on the whole, very different from those of their modern-day counterparts, historical understanding – at even the most superficial level – will elude us. (I should conjoin a caveat here. Purist historiography is *not* mistaken in its assumption that philosophy has been a fertile source for methodological ideas; that goes without saying. Where the purist approach becomes pernicious is in its emphasis on the exclusivity and primacy of the philosophical roots of methodology.)

My tactics, in seeking to make these claims plausible, will be these: I shall begin by putting forward several general theses about the historical character of philosophy of science; these are meant to outline an alternative to the purist's historiography of methodology; I shall then turn to discuss at length a few specific historical examples which will, I hope, serve as test cases for measuring the relative historical fecundity of the two historiographic models.

Looking first at the general theses, I would propose a *pragmatic, a posteriori* model for history. Its central tenets would be these:

(1) That the historically original and influential contributions to methodology have come *primarily* from working scientists and only secondarily from philosophers.

(2) That the epistemological theories of an epoch have generally been parasitic upon the philosophies of science of that epoch rather than *vice versa*, i.e. that the methodological ideas of working scientists have often been the source of major epistemological theories.

(3) That the traditional role for the philosopher of science has been predominately descriptive, explicative and legitimative, rather than normative; his avowed aim has been to make explicit what is already implicit in the best scientific examples.

(4) That the reception of methodological doctrines, even by the 'great' philosophers, has been determined more by the capacity of those doctrines to legitimate a preferred scientific theory than by their strictly philosophical merits.

Putting it otherwise: new or innovative methodological ideas have generally not emerged, nor have old ones been abandoned, as the result of an internal, dialectical counterpoint between philosophical beliefs; neither is it the case that the waxing and waning of methodological doctrines can be related neatly to their epistemic credentials. Rather, it is shifting *scientific* beliefs which have been chiefly responsible for the major doctrinal shifts within the philosophy of science.

With these two alternative models in mind, let me turn to discuss at some length an important episode in the history of the philosophy of science which I have been studying for some years.

III. THE METHOD OF HYPOTHESIS DURING THE ENLIGHTENMENT

It has been known for a long time that the beliefs of philosophers of science about the nature of scientific inference underwent a profound shift between the time of Newton and Mill on a large number of fronts. Probably most prominent here were the fortunes of the hypothetico-deductive method (or, as it was less clumsily called at the time, 'the method of hypothesis'). Commonly espoused in the middle of the 17th century by Descartes, Boyle, Hooke, Huygens and the Port-Royal logicians, the method of hypothesis fell into disfavor by the 1720s and 1730s. Few scientists and virtually no philosophers of science had any use for hypothetical inference. Knowing full well the fallacy of affirming the consequent, and its implications for the unreliability of the method of hypothesis, most scientists and epistemologists accepted the Baconian-Newtonian view that the only legitimate method for science was the gradual accumulation of general laws by slow and cautious enumerative inductive methods. Virtually every preface to major scientific works in this period included a condemnation of hypotheses and a panegyric for induction. Boerhaave, Musschenbroek, 'sGravesande, Keill, Pemberton, Voltaire, Maclaurin, Priestley, d'Alembert, Euler, and Maupertuis were only a few of the natural philosophers who argued that science could proceed without hypotheses, and without need of that sort of experimental verification of predictions which had been the hallmark of the hypothetical method since antiquity. As a contemporary noted, 'The [natural] philosophers of the present age hold hypotheses in vile esteem'.[2] Philosophers of science and epistemologists, were, if anything, even more enthusiastic in their condemnation of hypothetical inference. Whether we look to Reid's *Inquiries*, to Hume's *Treatise*, to Condillac's *Traité des Systèmes*, to Diderot's *Discours préliminaire*, or even to Kant's first *Critique*, the philosopher's refrain is the same: the method of hypothesis is fraught with difficulties; there are alternative methods of scientific inference, generally thought to be inductive and analogical ones, which alone can generate reliable knowledge. As Thomas Reid put it in 1785: 'The world has been so long befooled by hypotheses in all parts of philosophy, that is is of the utmost consequence ··· [for] progress in real knowledge to treat them with just contempt ···'.[3] The ardor with which 18th-century epistemologists repudiated the hypothetical method and endorsed the Newtonian inductive one is, in itself, plausible testimony to the impact of scientific archetypes on epistemology. But that is not the part of the story I want to investigate here. What interests me, rather, is a slightly later part of the tale; for *the self-same method of hypothesis which was so widely condemned by the 18th-century epistemologists and philosophers of science was, three generations later, to be resurrected and to displace the very method of induction which the philosophers and scientists of the Englightenment had set such store by.* If, for instance, we jump ahead to the 1820s and 1830s, philosophers of science such as Comte, Bernard, Herschel, Apelt, Whewell, Dugald Stewart, and even Mill were prepared to concede that the method of hypothesis had a vital role to play in scientific inference. And, with the exception of Mill, all these thinkers were prepared to acknowledge that the method of hypothesis was in fact *more central* to scientific inquiry than enumerative induction.

This about-turn, which effectively constitutes the emergence of philosophy of science as we know it today, is clearly of great historical importance. Explaining why and how the hypothetico-deductive method – which had been roundly condemned in the 18th century – came into prominence again, should presumably constitute one of the core areas of inquiry for the historian of the philosophy of science. Sadly, nothing could be further from the case. To the best of my knowledge, no scholar has been willing to confront this issue head on. Although several have noted that the process occurred, no one has offered an explanation for it. The avoidance of this historical puzzle cannot be mere indifference. It has been common knowledge for well over a century that the method of hypothesis waned and waxed between 1720 and 1840. One can only conjecture that the puzzle has not been solved because the confines within which most historians have worked do not allow of any cogent solution. They cannot do so because the purist historiography only allows one to explain changes within the history of methodology in terms of new philosophical doctrines and arguments. In the case at hand, these are unavailable because the logical and epistemological arguments which Stewart, Herschel and Whewell give for the method of hypothesis do not differ substantially from the kind of arguments which its 17th-century partisans had articulated. Indeed, the 19th-century criticisms of induction are, on the whole, no more than variations of criticisms of inductive inference which were well known in both the 17th and 18th centuries. Since no new philosophical arguments can be found which would explain the re-emergence of the method of hypothesis, or the repudiation of enumerative induction, the unstated assumption seems to be that the explanation of its fortunes is of no philosophical interest or significance and therefore a purely sociological question.

I want seriously to question that assumption. I believe that there is a straightforward intellectual explanation for the changing philosophical attitudes to the method of hypothesis during the period in question. Although we must look outside of philosophy proper to find the answer, the answer itself tells us something very signficant about philosophy itself, and about the ways in which philosophical beliefs are conditioned by changes in science.

The chief cause for the shift in attitudes of philosophers towards the hypothetico-deductive methodology was a prior shift in attitudes on the part of certain *scientists* towards the method of hypothesis. That latter shift, in turn, was determined by the changing character of physical theory itself, and by the tensions created when new modes of scientific theorizing ran counter to the inductivist orthodoxy prevalent among scientists and philosophers of science. I want to spell this process out in some detail.

For some fifty years after the triumph of Newton's *Principia*, both scientists and philosophers sought to draw the appropriate morals from the Newtonian success. As read by his immediate successors, Newton's achievement depended upon the eschewal of hypothetical reasoning and the rigid adherence to inductive generalization. For well over two generations, scientists sought to develop theories which could be regarded as straightforward inductive generalizations from the experimental data. Whether we look to the work of Hales, Boerhaave or Cotes, we see an effort to construct a purely observational physics, chemistry and biology whose ontology is immediately relatable to the data of experience.

By the 1740s and 1750s, however, scientists were discovering that many areas of inquiry did not readily lend themselves to such an approach. As a result, a number of scientists – and philosophers – began developing theories which, in the nature of the case, could not conceivably have been arrived at by enumerative induction. Franklin's fluid theory of electricity, the vibratory theory of heat, the Buffonian theory of organic molecules, phlogiston chemistry, these are but a small sample of the growing set of theories in the middle of the 18th century which hypothesized unobservable entities in order to explain observable processes. Among the most controversial of these theories were the chemical and gravitational theories of Georges LeSage, the neurophysiological theories of David Hartley, and the general matter theory of Roger Boscovich. Although working completely independently of one another, and differing over many substantive questions, these three thinkers – LeSage, Hartley, and Boscovich – have one very important characteristic in common: they quickly came to realize that the types of theories they were promulgating could not possibly be justified within the framework of an inductivist philosophy of science. Each of these thinkers found that his *scientific* theories, when once publicized, received widespread criticism, not because of their scientific merits or demerits, but rather because of their alleged epistemic and methodological deficiencies.

Against Hartley, it was claimed that his theory about aetherial fluids in the nervous system was but one of many hypotheses, between which only an arbitrary choice could be made.[4] Against Boscovich, it was argued that he could get no direct evidence that the forces around particles were alternately attractive and repulsive at the microscopic distances where contact, cohesion, and chemical change occurred. Against LeSage, critics contended that his theory of 'ultramondane corpuscles' (corpuscles whose motion and impact explained gravitational attraction) could not be inductively inferred from experiment.

Clearly, what confronted all these scientists was a manifest conflict between the accepted canons of scientific inference and the types of theories they were constructing. There was simply no way to reconcile an inductivist methodology and a sensationalist epistemology with such highly speculative theories about micro-structure. Their choice was a difficult one: either abandon micro-theorizing altogether (as their inductive critics insisted) or else develop an alternative epistemology and methodology of science which would provide philosophical legitimation for theories which lacked an inductivist warrant. All three in our trio chose the latter alternative. Boscovich insisted that the method of hypothesis is 'the method best adapted to physics' and that, in many cases, it is only by means of conjecture followed by verification that 'we are enabled to conjecture or divine the path of truth'.[5] Hartley, in a lengthy chapter on methodology in his *Observations on Man*, asserted that the methods of induction must be supplemented by various hypothetical methods if we are ever to accelerate the acquisition of knowledge beyond a snail's pace.[6]

The most explicit defense of hypothesis, however, came from Georges LeSage, whose theory had been severely attacked.[7] Euler, for instance, had said that it was better to remain ignorant 'que de recourir á des hypothéses di étranges'.[8] The French astronomer Bailly had insisted, in good inductivist fashion, that science should limit itself to those 'lois qu'elle nous a manifestées',[9] and avoid conjecturing about what we cannot directly observe. There are, LeSage laments, 'almost universal prejudices' that hypothetical reasoning from the observed to the unobserved is impossible, and that induction and analogy are the only legitimate routes to truth.[10] He points out that his own theory was being widely dismissed because 'mon explication ne peut être qu'une hypothèse'.[11]

Confronted with such attacks, LeSage was forced to play the epistemologist. In several later works, but especially in an early treatise on the method of hypothesis, written for the French *Encyclopédie*, LeSage began the counterattack. In brief, his strategy was two-fold: first, to establish the epistemic credentials of the method of hypothesis; second, to point up a number of weaknesess in the dominant inductive and analogical accounts of methodology. LeSage, in short, agreed with his critics that his theory was indeed an hypothesis; but unlike them, he sought to show that it was none the worse for that.

He grants immediately that the method of hypothesis and verification can rarely establish the truth of any general conclusions. But then, as he points out, induction and analogy are also inconclusive. What we must aim at in these matters is high probability and LeSage indicates circumstances under which we are entitled to assert well-confirmed hypotheses with confidence. He goes on to point out that the great Issac Newton, for all his professed inductivism, extensively utilized the method of hypothesis and that it is 'solely to [the method of] hypothesis, without any element of [induction or] analogy that we ··· owe the great discovery of the three laws which govern the celestial bodies'.[12] Generalizing this point, LeSage argued that there is an element of conjecture or hypothesis in every inductive inference which goes beyond its premises, which all except so-called perfect inductions do. He spent much of the next half century trying to resurrect the method of hypothesis, conceived along these lines.[13]

Although this is not the place to present it, the evidence is as unambiguous as evidence can be that LeSage's lengthy and persistent espousal of the method of hypothesis was conditioned by his prior scientific committments and by the epistemic criticism directed against his scientific theories. Similar, if more circumstantial, claims can also be made for Hartley and for Boscovich.[14]

But the story does not end here, for there is the larger question as to how the method of hypothesis, here propounded by a tiny handful of beleaguered scientists, worked its way into philosophical orthodoxy. There are a number of pieces to the larger puzzle. Chief among them are these: Jean Senebier, a French philosopher-scientist best known for his work on photo-synthesis, writes an influential three volume work in 1802 on scientific method which, following LeSage, endorses the method of hypothesis and accumulates further evidence for the wide-spread use, among the best physical theories, of the method of hypothesis.[15] Shortly thereafter, Pierre Prevost, founder of the theory of heat exchange, publishes a posthumous collection of LeSage's essays on hypothesis and himself writes a book in 1804 on philosophy of science in which prominence is given to the method of hypothesis.[16] In Scotland, Dugald Stewart – the leading English-language philosopher at the turn of the century – repudiates the trenchant inductivism of his mentor Reid and, after an explicit discussion of LeSage, Hartley and Boscovich, warmly endorses the method of hypothesis, since 'it has probably been this way that most [scientific] discoveries have been made'.[17] As Olson has recently shown, Stewart's discussion of the philosophy of science provided the framework from which Herschel and Whewell, prominent proponents of the method of hypothesis, drew heavily.[18] (One might add that the scientific committments of Herschel and Whewell to the wave theory of light also seem to have had much to do with their espousal of the method of hypothesis.) In the case of each of these writers – from Hartley and LeSage through Prevost and Stewart to Herschel and Whewell – the explicit and over-riding consideration in their endorsement of the hypothetico-deductive method (as against induction or analogy) was *that the sciences of their time required such a method.* Although familiar philosophical arguments in favor of the method of hypothesis are routinely rehearsed, and a few innovations are introduced, the primary factor which justifies that method – in the view of its proponents – is the raw fact that it is being used in the most successful sciences. Even Mill, who would dearly have loved to dispense with the method of hypothesis, felt constrained – given its widespread use in the sciences – to find a logical rationale for it.[19]

IV. CONCLUSION

Hence, it seems impossible to make any historical sense of this episode by looking at scientific epistemology as autonomous and self-contained. Sadly, but predictably given the predominance of the purist model, ones looks in vain in the scholarly literature for any rival to this account of the emergence of the hypothetico-deductive method. Le-Sage, Hartley, Lambert, Prevost, are never even mentioned in histories of the philosophy of science. Although Herschel and Whewell are often discussed, no attention is paid to their intellectual ancestry nor to the scientific interests which shaped their treatment of hypotheses. Instead, the standard literature leads us by the hand from Newton to Hume to Kant to Whewell, with no hint that most of the methodological-theorizing on this vital topic was taking place outside the mainstream – or perhaps I should say outside the *presumed* mainstream – of epistemology. But is this case typical, or is it just the exception which proves the proverbial rule?

Upon the answer to that question hangs the fate of the traditional, purist historiography of method. I am inclined to think, judging by some of the best recent scholarly studies, that we are dealing with a common phenomenon. Although such studies are still very much in the minority, it is worth considering some of the recent conclusions: Mittelstrass and others, following the lead of Duhem, have shown that instrumentalism has its historical roots in debates within ancient and renaissance astronomy;[20] Mandelbaum has shown the intimate connections between atomism and the doctrine of primary-secondary qualities;[21] Edelstein and others have shown the connections between

Pyrrhonic scepticism and Greek medicine; Buchdahl and Sabra have explored the impact of optics on epistemology in the 17th century.[22] Recent studies by others have shown that it was ultimately the triumph of atomic-molecular physics in the early years of this century that eventually tipped the balance in the philosophical debate between realists and instrumentalists.[23] The historical record leaves no doubt that the historical fortunes of many philosophies of science have been closely intertwined with the fortunes of those scientific theories upon which the philosophies were modelled.

What all these interconnections suggest is that traditionally the philosophy of science and large parts of epistemology have been modelled on, influenced by, and devised to legitimate, certain preferred or priviliged forms of scientific activity. And why, in a sense, should we have imagined otherwise?

In exploring the nature of science or knowledge, it is perfectly natural that methodologists and epistemologists should have selected as their explicanda the best available examples of science or knowledge. The presumption, so familiar in the 20th century, that general methodology can be entirely normative and purely *a priori*, and thus not parasitic upon any specific examples of knowledge was not a conception cherished by our predecessors. They were, in general, prepared to concede, even proud of the fact, that the sciences were justificationally prior to the philosophy of science and that the epistemologist's aim was, in Locke's language, to serve not as judge of, but as underlaborer to, his scientific contemporary. Their task, as they perceived it, was not to *prescribe* what methods the scientist should follow; but to *describe* the best methods being used in scientific *praxis.*

But we would be missing the point to imagine that all that is being claimed here is that scientific theories somehow have inspired or motivated methodological doctrines, and that the latter, once invented, acquired a life of their own and had a temporal career entirely independent of the former. We need to realize that scientific theories not only inspire new theories of methodology, they also – in a curious sense – serve *to justify those methodologies.* For instance, the success of Newton's physics was thought to sanction Newton's rules of reasoning; Lyell's geological theory was cited as grounds for accepting methodological uniformitarianism; the kinetic theory of gases and

Brownian motion were thought to legitimate epistemological realism; these are but a few examples of a very common phenomenon.

Given this close symbiotic relation between science and its philosophy, it is entirely natural that when changes take place among the theories regarded as archetypically scientific so, too, should the assessment of methodological doctrines shift. And this is really the nub of the whole matter. Confronted by a range of equally convincing or unconvincing philosophies of science, *our predecessors often looked to the sciences of their time as the appropriate laboratory for evaluating competing philosophies.* They insisted that methodological principles should be 'empirically' tested by seeing whether they could be used to legitimate those scientific theories which were taken to be the best available examples of knowledge. If many of us no longer see the situation in that form, we should nonetheless be imaginative enough historically to be mindful of such issues when we talk about the past. Indeed, whatever the relative merits of these two divergent views about the relation of philosophy of science and the sciences, it is clear that the *historian* must be more prepared than he has been to look beyond the narrow confines of the purely *philosophical* if he is to make any sense of the history of methodology.

My thesis thus far has amounted to this: that one cannot understand the history of methodology without looking carefully at the historical evolution of those sciences upon which methodology has traditionally been parasitic. There is an important corollary to this thesis which has already been made explicit but not discussed; to wit, that precisely because the history of methodology is more than just a branch of the history of epistemology, we must abandon the view, presupposed by so much recent scholarship, that the thinkers who loom large in the history of epistemology should *a forteriori* be accorded pride of place in the history of methodology. It simply is false to assume that there is a neat intersection between the two groups.

Let me draw my examples again from the middle and late 18th century. The prominent epistemologists, of course, were Hume and Reid in Britain, Condillac and Condorcet in France, Wolff and Kant in Germany. Few of these thinkers would figure prominently on a list of original and influential methodologists for the same period. If we would learn who contributed most to such problems as the methods of induction, techniques of hypothesis evaluation, the articulation of various experimental and observational methods, the application of probability theory to scientific inference, the evaluation of claims about theoretical entities and the like, we must look to a very different constellation of figures. We must look to Hartley, LeSage, and Lambert, who between them elaborated the hypothetico-deductive method at a time when the major philosophers of the day had nothing to say in its favor. We must look to Bernoulli, Mendelsohn, Laplace and d'Alembert for discussions of the logic of probability. We must look to Pierre Prevost, and to Jean Senebier's classic text on philosophy of science to see the most sophisticated treatment of the various experimental methods and their logical foundations.

The very fact that most of these latter figures are virtually unheard of nowadays, at least within the history of methodology, is further testimony to the extent to which we have allowed our scholarly image of the history of the philosophy of science to be warped by the naive subsumption of methodology under epistemology. If, to return the language of my title, we would learn something about the sources of modern methodology, we must look to a different set of issues and a different set of thinkers than those that have customarily monopolized our attention.

University of Pittsburgh

NOTES

* I am indebted to the National Science Foundation and the University of Pittsburgh for support of portions of the research for this project.
[1] For a guide to many of these studies, see my 'Theories of Scientific Method from Plato to Mach: A Bibliographic Review', *History of Science* 7 (1968) 1–63.
[2] Benjamin Martin, *A Philosophical Grammar* (London, 1748), p. 19. Similar sentiments are expressed a year later by Condillac; cf. his *Oeuvres* (Paris, 1798), vol. 2, pp. 327ff.
[3] T. Reid, *Works* (ed. W. Hamilton, 6th ed., Edinburgh, 1863), vol. 1, p. 236. For further discussion of this background, see L. Laudan, 'Thomas Reid and the Newtonian Turn of British Methodological Thought', in R. Butts and J. David (eds.), *The Methodological Heritage of Newton* (Toronto, 1970), 103–31.
[4] Most contentious in Hartley's system was his effort to provide a neuro-physiological foundation for the Lockean 'associationist' psychology by postulating an aetherial fluid which filled the nerves.

[5] From Boscovich's *De Solis a Lunae Defectibus* (1760). Quoted from, and translated by D. Stewart in his *Collected Works* (ed. by W. Hamilton, Edinburgh, 1854–60), vol. 2, p. 212.

[6] See especially Hartley's *Observations on Man: His Frame, His Duty and His Expectations* (London, 1749), vol. 1, pp. 341–51.

[7] For a discussion of LeSage's physics, see S. Aronson, 'The Gravitational Theory of George-Louis LeSage'. *The Natural Philosopher* 3 (1964) 51–74; for a brief discussion of that theory's philosophical significance see L. Laudan, 'George-Louis LeSage: A Case Study in the Interaction of Physics and Philosophy', in *Akten des II. Internationalen Leibniz-Kongresses, Hanover, 17–22. Juli 1972* (Wiesbaden, 1974), vol. 2, pp. 241–52.

[8] From a letter published in *Notice de la Vie et des Ecrits de George-Louis LeSage* (ed. P. Prevost, Génève, 1805), p. 390.

[9] *Ibid.*, p. 300.

[10] *Ibid.*, p. 265.

[11] *Ibid.*, pp. 464–65.

[12] This quotation is from LeSage's 'Premier Mémoire sur la Méthode d'Hypothèse', published posthumously in P. Prevost's *Essais de Philosophie* (Paris, 1804), vol. 2, para. 23.

[13] For references to these later works, see L. Laudan, *op. cit.*, note 7.

[14] I am now doing a comparative study on the reception of the theories of Hartley, Boscovich, LeSage and Lambert.

[15] See his *L'Art d'Observer* (2 vols. Génève, 1775), expanded to the three-volume *Essai sur l'Art d'Observer et de Faire des Expériences* (Génève, 1802). Senebier, incidentally, was LeSage's successor as Director of the Geneva library.

[16] See P. Prevost, *op. cit.*, note 12.

[17] D. Stewart, *op. cit.*, note 5, vol. II, p. 301. (Cf. also *ibid.*, pp. 307–308.)

[18] See Richard Olsen's very interesting study, *Scottish Philosophy and British Physics, 1750–1880* (Princeton, 1975).

[19] Cf. Mill's chapter on hypotheses in the *System of Logic*.

[20] See J. Mittelstrass, *Die Rettung der Phänomene* (Berlin, 1962).

[21] See M. Mandelbaum, *Philosophy, Science and Sense Perception* (Baltimore, 1964).

[22] See A. I. Sabra, *Theories of Light from Descartes to Newton* (London, 1967); and G. Buchdahl's numerous studies of Descartes.

[23] For a brief discussion on this issue, see L. Laudan, 'The Methodological Foundations of Mach's Anti Atomism and their Historical Roots', in P. Machamer and R. Turnbull (eds.), *Motion and Time, Space and Matter* (Columbus, 1976), pp. 390–417.

L. A. MARKOVA

DIFFICULTIES IN THE HISTORIOGRAPHY
OF SCIENCE

Nowadays we read rather often about a revolutionary situation in the modern historiography of science. I shall try to outline the main difficulties and the unresolved problems which give rise to such opinions. Even a superficial acquaintance with modern works on the history and philosophy of science gives us an opportunity to notice three difficulties, each of them attracting intense attention from investigators. All these difficulties are connected with the introduction of the idea of discontinuity in the history of science in opposition to the positivistic interpretation of the history of science as continuous and cumulative.[1]

First of all, the role of social factors in the development of science is being vividly discussed. The question is put approximately in the following way. Science is engendered by human society and exists in it. The speed and direction of its development, and the range of problems considered depend to a large extent on social conditions. Scientific theories are founded by a scientist as a man belonging to a definite epoch, country and class, and these peculiarities of his life define in many respects the success of his scientific activity. The social structure of science itself, the functioning of scientific institutes, and the forms of personal contact between representatives of the same scientific community or of different ones – all this directly influences the character of scientific work. Connections of such a kind are obvious, and nobody denies them. They are investigated in detail in all their diversity by the historians of science of the externalist trend (R. Merton, E. Zilsel, J. Needham, G. Sarton etc.) and in the science of science works.

Meanwhile, in spite of the obvious character of these relations, a gap between the social and the logical, between the social relations and the internal logic of scientific knowledge always remains. In this the first difficulty consists. It was always difficult to penetrate the immanent process of the development of scientific ideas from the point of view of externalism. In connection with this a difficult question is often put to

Butts and Hintikka (eds.), Historical and Philosophical Dimensions of Logic, Methodology and Philosophy of Science,
21–30.
Copyright © 1977 by D. Reidel Publishing Company, Dordrecht-Holland. All Rights Reserved.

the representatives of the externalist trend: if you cannot explain by your own means the development of scientific ideas, for instance, to explain how relativistic physics came into being from Newtonian physics, what do you have in common with the history of science? Do the external factors play any other role in the emergence of scientific theories than their role, recognized by everybody, as stimulators or hindrances? From the point of view of externalism it is possible to write a history of science as a social institution, but this history exists only side by side with the history of ideas, not coinciding with it. As M. Finocchiaro writes in his book, social factors can be a cause of the social aspects of science, no more. (Finocchiaro, 1973.) The gap between social factors and scientific knowledge turns into a gap between the social history of science as a history of social institutions and the intellectual history of science as a history of ideas. Two histories – two sciences. However, in practice, in real life, this gap is constantly bridged, and the task is to understand theoretically the transition from the social to the logical.

The second essential difficulty which historians of science encounter is connected with the interrelations of a theory with physical reality, or, in other words, of a theory with a fact. In what way can the facts of physical reality influence theory, corroborate or refute it, either through everyday experience, through the capacity of our organs of sense to perceive the external world, or through a scientific experiment? Or, otherwise expressed, in what logical forms is it possible to reproduce in a theory the lack of correspondence between a theory and its object? In the history of philosophy the attempts to get an answer to this question were made usually in the course of the investigation of the problem of induction. Among the historians of science A. Koyré was one of the first to show the whole complexity of the question about the interrelation between scientific knowledge and physical reality, and its relevance for the tasks of the history of science. In opposition to positivism Koyré underlined the impossibility of a direct inference from experience to scientific knowledge. To his mind, the world of science and the qualitative world of perceptions in which we live and die, are separated in the sphere of theory by a gap, though in practice they get in touch with one another every day. Investigating the role of experiment in modern science Koyré concludes that imaginary experiment "plays a part intermediate between pure thought and tangible experiment" (Koyré, 1968, p. 82).

To my mind, the significance of Koyré's ideas for the historiography of science consists, first of all, in his introduction of the idea of discontinuity into the history of science. He is against the positivistic point of view which claims the possibility of inferring scientific knowledge directly from experience and from social demands, and also he was against the cumulative interpretation of the history of scientific ideas. Koyré saw a problem where for positivistic historians of science everything was clear and unquestionable. Koyré is a leader of the internalist school in the historiography of science, but this doesn't mean at all that he rejected rashly any influence of social factors on the development of science. He just couldn't see the possibility of theoretically understanding this influence, which is why he usually refused to consider this aspect of scientific development.

We have come directly to the third difficulty of the historiography of science, namely, to the difficulty connected with the interpretation of scientific revolutions.[2] In this case many historians and philosophers of science question the very possibility of a rational interpretation of the process of the emergence of new knowledge. Problems connected with this have come forward especially clearly in the course of discussion about Kuhn's book *The Structure of Scientific Revolutions* (Kuhn, 1962).

The situation in the historiography of science thus briefly outlined has found a peculiar expression in the conception of the three worlds of K. Popper (Popper, 1972). If one may put it that way, Popper revealed, or brought to light, a boundary to which history and philosophy of science approach: the third world of ideas, developing through a continuous series of revolutions irrespective of the world of man and irrespective of the world of physical reality. All three lines of demarcation or gaps which we have spoken of are present here: the gap between scientific knowledge and social conditions, the gap between scientific knowledge and physical reality, and the gap between old and new theories. As it seems to me, the main efforts of contemporary historians and philosophers of science are directed, explicitly or implicitly, to overcoming these gaps, to the construction of rational bridges over them.[3]

Now I shall try to show that the same understanding of scientific knowledge lies at the bottom of all difficulties connected with the overcoming of gaps of each of the three types.

Let me remind you of two different aspects of scientific knowledge. They are knowledge put into the form of a scientific theory with the help of mathematical or other technical tools, and knowledge in the process of coming into being ('public' and 'private' aspects in the terminology of G. Holton). Considered in its first aspect, scientific knowledge is knowledge 'for others'. It can be used by other scientists who are not its inventors and by engineers and technicians in production. Such a relation with industry is usually illustrated by the production of things useful for man. Inverse relation by means of social orders is a relation with the results of knowledge as well. Social order is an order which assigns a definite product to the scientific work. The products of scientific work are assimilated to the products of material production.

Such a relation between science and society has been the subject of special attention on the part of philosophers since the time of F. Bacon, who was among the first to emphasize the active character of science, its capability for influencing nature and its place in the service of man. Hardly anyone has laid more stress on the utilitarian aspect of science than the early positivists (Comte, Spencer, Mill). As this view of science directed attention first of all to its results, the further development of positivism led, on the one hand, to a refusal of its representatives to analyze the problems of development in the framework of logical positivism, and on the other hand, to an externalist understanding of the history of science, which is based on the belief that the causes of the development of scientific ideas lie exclusively in the needs of society.

The very same interpretation of scientific knowledge as a constellation of results hinders the resolution of the problem of finding the logical forms which would reproduce in a theory the lack of correspondence between a theory and its object. In this case externalist relations are also present in the reasoning of philosophers and historians of science. External material nature is recognized as an object of scientific investigation, but this recognition remains only as a simple premise. The relation of a theory to nature either doesn't play any role in a

conception, or this relation is interpreted exclusively from the externalist point of view: a crucial experiment is considered often as refuting an
old theory and confirming a new one. In this case the same difficulties
emerge which we have talked about in connection with the externalist
approach to the analysis of the relations between science and society.
The question is: can any phenomena of the external, material world
play a role in scientific development other than as an external stimulus
to prompt the emergence of a new theory and the rejection of an old
one? Can they influence the logic of scientific reasoning?

When engaged in the interpretation of scientific revolutions, historians of science deal with the constellation of received results as well;
only in this case these results are considered, not in their relation to
external social factors, nor in their relation to the facts of physical
reality, but in their relation to one another. Internalist historians
emphasize a genuinely characteristic aspect of modern science: a
scientist operates in his everyday work with knowledge at his disposal
and with safe, unproblematic tools for the achievement of definite
results. This aspect of scientific activity is brilliantly represented by T.
Kuhn in his analysis of normal science (Kuhn, 1962). He brings his
investigation to a logical limit, and then it becomes clear that scientific
activity, understood as dealing solely with received results, is not
compatible with creative thinking. When concerned with products the
investigator loses all interest in the previous process of the emergence
of knowledge and in the scientist as a subject of creative work.
Theories move, replace one another in Popper's third world of ideas,
and for the understanding of their development it is not necessary to
leave the framework of this world. External relations are introduced
into the immanent analysis of scientific development.

Using the results of thinking in industry permits one to manufacture
material things. Normal activity, consisting in operation with these
results, creates a type of scientist whose work can be identified with the
process of material production when, in advance, the result is known
entirely (in production) or almost entirely (in science), and there are on
hand the tools for its achievement, safe and tested in the same
circumstances (in production) or in analogous circumstances (in science). Such a kind of scientific activity stamps not only the mode of
scientific reasoning but also a modern way of thinking in general. To

no small degree this activity underlies antiscientific sentiments and reproaches directed against science, that it wastes man, turns him into a mechanism, an automaton deprived of emotions and creative abilities.

The logic of the third world of ideas, adjusted for the study of the development of results of scientific investigation, does not accommodate the exploration of creativity. After a number of attempts to give a rational interpretation of scientific revolutions, we have had confessions by authors of these attempts to failures in their enterprise. It is in this that the second main difficulty of the historiography of science consists.

In modern science (after Galileo) usually the products of scientific knowledge and the mode of their circulation, both in science and outside of it, came forward in the end, but creative processes always remained the source of the emergence of new theories, conceptions, and ideas. It is true that these processes were not of great interest and rarely became the subject of special exploration. Nowadays, when they attract attention, they seem to be something absolutely alien to scientific activity as it is commonly understood; this situation is reflected in Kuhn's book as well. In his opinion a normal activity in science proceeds in quite a different way from a revolutionary one. Meanwhile it is necessary to take into account the normal activity for the understanding of the historical specificity of scientific creativity. The relation between them becomes apparent, for instance, in the fact that a scientist, after he has attained some new results, at once hurries to forget the way which led him to them. The main task for him is then to shape his results 'for others', in such a way that they can understand and use them.

Such a purpose to the delivery of products influences, surely, the creativity itself; though by definition creative work cannot operate with the results of thinking. It is not a reproduction, in contrast to the activity in industry where the process of production is rigidly determined by the end; it is necessary to make such and such a thing and to use such and such a means of production. In this case the end is a law (K. Marx). In creative work we change our point of departure (axioms, postulates, theories) and the end as well. By the way, this is brilliantly demonstrated by I. Lakatos in his book 'Proofs and Refutations' (Lakatos, 1963–4).

Discussions of T. Kuhn's book demonstrate clearly the shift of interest from the process of the emergence of new knowledge to the procedure of the choice between theories which are already available to us. The very terminology used by participants in these discussions testifies to the shift of interest to the analysis of results already achieved by investigations. It is the problem of 'choice' or of 'preference' of one theory over another, or the problem of 'competition' between an old theory and a new one that is being discussed.

I want to emphasize that externalism, conceived of as a way of explaining historical events through their relations to something external to it, does not coincide with the sociological interpretation of the history of science. In immanent conceptions, externalist relations are present to no less an extent; they are present in the form of interaction between theories as results of scientific investigations, each of them being external to all others.

Nowadays reaction against positivism among historians and philosophers of science manifests itself in their claim to study, first of all, the growth or change of scientific knowledge, not the structure of a theory, and to consider, secondly, the growth and development of scientific knowledge as accomplished through revolutions, not cumulatively and gradually. And indeed, in contemporary antipositivistic conceptions, the sequence of many theories, competing with and replacing one another, is considered rather than the structure of a theory. But again these theories are present in conceptions as already founded, not in the process of their emergence. As to scientific revolutions, their rational interpretation is hampered very much by the fact that the positivistic thesis about the inferring of new knowledge out of the old is rejected, and as a result, historians and philosophers are unable to establish any rational relation between two theories following one after another and considered as products, not in the process of their emergence. Between them an unbridged gap arises.

In my opinion, historians and philosophers of science now encounter a very important task. After the rejection of the positivistic idea of cumulative development, it is necessary to take the next step, which is to refuse to consider the history of science as a replacement of one theory by another and to begin to analyse scientific creativity, to analyse theories in the process of coming into being. A precondition for this step is present, and a very serious one. I mean the inclusion of philosophy into scientific knowledge, which was accomplished for the first time among historians of science by E. Burtt in his book *Metaphysical Foundations of Modern Physics* (Burtt, 1925). It was a serious attack on the most important foundation of positivism, all forms of which from Comte to Carnap are unified by separation of scientific knowledge from philosophy. The inclusion of philosophy in scientific knowledge meant the introduction there of questions of the relation of thinking to being, both to social and physical being. An opportunity now appears to consider the social aspect of science through analysis of its inner structure, and not through its relation to something external to it.

In the contemporary historiography of science this opportunity is realized in so far as the concept of thought-experiment is elaborated. It is in the course of a thought-experiment that theoretical interaction of the ideal object of scientific investigation, its interpretation, and the subject of scientific activity is accomplished. The subject of the activity, either as an individual scientist, or as a scientific community, embodies in itself, as in focus, social characteristics; and an ideal object represents physical reality inside a theory. The ideal object is a direct subject of investigation, though indirectly in the end science studies physical, material reality. Physical reality exists for a theory not only as a knowledge, as what we know, but also as a lack of knowledge, as what we do not know. For example, in Aristotelian physics projectile motion was an instance of this 'lack of knowledge', the theory itself determined what was unknown. The determination of such inconsistency between theory and its object is one of the philosophical functions of theory.

As an illustration of these considerations, let us recall Koyré's characterization of the concept of movement in modern mechanics: "In modern science, as well we know, motion is considered as purely geometrical translation from one point to another. Motion, therefore, in no way affects the body which is endowed with it; to be in motion or to be at rest does not make any difference to, or produce a change in, the body, whether in motion or at rest. The body, as such, is utterly indifferent to both." (Koyré, 1968, p. 4.)

As it seems to me, it is from this understanding of movement that the contemporary mode of explanation of the growth of scientific theories originates: theories are in motion only in relation to something else – to some other theory, or to some social factors, or to some facts of physical reality; but this motion doesn't explain their inner logical structure. Such an externalist approach to the development of scientific knowledge reveals many interesting and very important aspects of the history of science, but 'the point of growth' of the historiography of science is not there now.

The externalist approach, which is characteristic in equal measure both to the externalist and the internalist trends in the historiography of science, is of no use for accomplishing the tasks and overcoming the difficulties discussed above. All three types of relations (between theory and social factors, between theory and the facts of physical reality, and between different theories) must be studied as peculiar to the inner structure of scientific theory in the process of its coming into existence – the direct object of investigation being thought-experiment, the core of a scientific revolution.

NOTES

[1] On the positivistic interpretation of the history of science see Gorsky, D. P. and Griasnov, B. S. (eds.), (1975).
[2] On this topic see Mikulinsky, S. R. (1964, 1972).
[3] An interesting attempt in this direction was made in the book of Bibler, V. S. (1975).

BIBLIOGRAPHY

Bibler, V. S.: 1975, *The Thinking as a Creativity*, Moscow (in Russian).
Burtt, E. A.: 1925, *The Metaphysical Foundations of Modern Physical Science*, London.
Gorsky, D. P. and Griasnov, B. S. (eds.): 1975, *Positivism and Science*, Moscow (in Russian).
Finocchiaro, M.: 1973, *History of Science as Explanation*, Detroit.
Koyré, A. (ed.): 1968, 'Galileo and the Scientific Revolution', in *Metaphysics and Measurement*, Harvard, 1968.
Koyré, A. (ed.): 1968, 'Galileo's Treatise de Motu Gravium: the Use and Abuse of Imaginary Experiment', in *Metaphysics and Measurement*, Harvard, 1968.
Kuhn, Th.: 1962, *The Structure of Scientific Revolutions*, Chicago.
Lakatos, I.: 1963–4, 'Proofs and Refutations', *The British Journal for the Philosophy of Science*, **14.**

Mikulinsky, S. R.: 1964, 'Methodological Problems of the History of Biology', *Voprosi Philosophii*, No. 12 (in Russian).
Mikulinsky, S. R. (ed.): 1972, 'Introduction', in *The History of Biology*, Moscow (in Russian).
Popper, K.: 1972, *Objective Knowledge*, Oxford.

LOGICAL, ONTOLOGICAL AND METHODOLOGICAL ASPECTS OF SCIENTIFIC REVOLUTIONS

INTRODUCTION

Scientific revolutions – from Copernicus to Einstein and Bohr – have been the subject of many studies, but I don't think the topic has been exhausted or even treated in the manner it deserves. Usually, the revolutionary elements are overrated and the conservative ones overlooked or underrated. The most notorious example of such an approach is Kuhn's (1962) conception of scientific evolution as a sequence of 'paradigms'. In contrast to such views I shall endeavour to show that the scientific revolutions of the past are highly conservative, indeed in more than one respect, and that this is one reason for their success. Thus, the modern historiological anarchists will find no support in the following pages. On the other hand, the conservatives will draw no comfort either. Not only are further revolutions in fundamental theory unavoidable, but in examining both the dialectics of continuity and discontinuity in previous revolutions and the present needs for theory change we shall find reasons to expect that future revolutions in fundamental theory will show quite novel features hardly fitting into the present schemes of theory change.

The dominating questions of our discussion will concern both the *nature* and the *extent* of changes in fundamental theory. By 'extent' I mean the question of how far a change in a fundamental theory affects the remaining theories or even the whole structure of physics. By 'nature of a change' I mean both the object of change and the way it has been changed. The primary object of change may be any of the three elements constituting a physical theory: mathematical substructure (M_{sub}), mathematical superstructure (M^{sup}), physical interpretation (PI) (semantics). Now the PI always implies some kind of ontology; hence change in fundamental theory may be connected with a

Butts and Hintikka (eds.), Historical and Philosophical Dimensions of Logic, Methodology and Philosophy of Science, 31–49.

change in ontology, but such a change would be an implied or secondary one, at least from the logical point of view, though it may be primary in the heuristics that led to the new theory.

As the ontological aspect of theory change has not been explicitly considered in my previous publications I shall start with some remarks on this subject. For the rest I shall assume that the reader is familiar with the contents of my book (Strauss, 1972), particularly with the chapters on the evolution of physics (Chaps. 1, 2) and intertheory relations (Chap's. 12, 15, 22) to which frequent reference will be made.

1. Ontic implications of a physical theory

By 'ontic implications of physical theory' I mean, roughly speaking, the sort of objects implied to exist by that theory when the theory is given in the usual (semi-axiomatized) standard form

$$(1) \qquad PT = M_{\text{sub}} \cap M^{\text{sub}} \cap PI.$$

To be sure, in a completely axiomatized theory the question of ontic implications does not arise as a problem because the existence of the objects concerned would be postulated. Yet complete axiomatization of a physical theory is a highly controversial undertaking since it can be carried out in many different ways. Moreover, different axiomatisations may even differ in their ontic commitments. Thus the kinematics of Special Relativity, which involves the velocity constant 'c', may be axiomatized with or without the postulate that objects with the velocity c (with respect to a fundamental frame in Minkowski spacetime) exist. Though empirically true, the named postulate is not required to establish the kinematics of SRT, and the mass point dynamics of SRT merely implies that particles of zero rest mass, if they exist, must have the velocity c. Since it seems to be good policy to keep the ontic commitments of a fundamental theory to a minimum the PI in (1) will always be assumed to be free of unnecessary ontological assumptions.

More useful, because less arbitrary, than axiomatisation is *formalisation*. Indeed, if formalisation is carried out according to the method ('Mathematics as Logical Syntax') already used by the author in 1938 (Strauss, 1972, Chap. 6) the formalisation procedure is practically unique. In essence the method consists in the introduction of physical predicates uniquely related to those mathematical symbols which represent primary properties of physical objects; in the formalised physical language the objects are then represented by individual variables or constants not occurring in the mathematical formalism $M_{sub} \cap M^{sup}$ but serving as arguments of the sentential functions (predicates). Quite obviously, the objects so represented can only be objects of which the property in question can be predicated. Thus, *it is the physical predicates of a theory which determine the ontic implications of that theory.*

Incidentally, if questions of logical structure are not involved we may use instead the simpler method of *semantic completion* of the mathematical notation used in the M^{sup} of the theory. Thus, instead of the usual expression for Newton's Second Law

$$\frac{d}{dt}(m\mathbf{v}) = \mathbf{F}$$

we would write

$$(\exists \Sigma)(\forall A)\left[\frac{d}{dt}(m_A v_A^\Sigma) = \mathbf{F}_A^\Sigma\right],$$

where 'A' means any point particle and 'Σ' a frame of reference. (As the example shows, semantic completion compels one to think about quantification – a very useful side effect!)

A special remark is required concerning field theories. The mathematical formalism (MF) of a classical field theory consists of a M_{sub} specifying the structure of a spacetime and the so-called field equations, i.e. partial differential equations for the field quantities which are functions over the specified spacetime. If now the distribution of field quantities is considered as a property of spacetime, the latter acquires an ontological status – contrary to both the Clark-Engels conception of space and time as forms of existence of matter and the so-called relational conception of space and time which goes back to Leibniz. The dilemma is resolved if we consider a distribution of field quantities, i.e. a solution of the field equations, as a *state* or property of the *field*, just as we take 'mass' to be a property of a *particle*. And just as there are particles with zero rest mass, there may be fields with zero field strengths everywhere – a conception corroborated by quantum field theory.

Thus, the material object implied to exist by a field theory is indeed a *field*, not a state of the field (which is a possible property of the field) nor a spacetime (which is not a material object, whatever else it may be). What we can say generally about a physical spacetime is merely that some of its *properties* reflect certain properties of the primary properties of physical matter; thus, the three-dimensionality of physical space reflects the property of the extensions of macroscopic bodies to be three-dimensional. The mathematical spacetimes used in physics certainly have more properties than can be related to the properties of primary properties of physical matter, but this does no harm as long as these additional properties do not conflict with the postulated properties of matter. After all, we always 'buy more' from the mathematical store than we actually need, perhaps because the thing we really need is not available, the notable exceptions being the algebraic structures.

2. THE GREAT REVOLUTIONS

2.1. *The Heroic Period: Change in Method*

Seven names may be selected to characterize what we like to call the heroic period: Copernicus (1473–1543), Tycho (1546–1601), Galileo (1564–1642), Kepler (1571–1630), Huygens (1629–1695), Hooke (1635–1703), and Newton (1642–1727). These men did not agree on all questions of theory, but they all subscribed to, and practiced, the new revolutionary method characterized by observation, experiment, and the use of mathematics, all subordinated to the maxim: *analysis has to precede synthesis.* It is this method which forms the true foundation of all subsequent progress in physical science. There is no need here to go into details, but a few remarks may be allowed on points usually overlooked or misrepresented.

(a) The specific change advocated by Copernicus can be described as a *change in the kinematic model* of the solar system. Leaving aside the affront to theological doctrine, the heliocentric model, taken by itself, can hardly be called revolutionary, quite apart from the fact that Aristarchus had invented it about seventeen hundred years before. Indeed, in the spacetime $E_3 \times T$, not questioned until 1905, all frames of reference are of equal rights, and the most a kinematic model can claim is maximum descriptive simplicity. It is only under *historical perspective* that the work of Copernicus can and must be called revolutionary because the heliocentric model was a necessary step towards a celestial *dynamics*. In fact, it anticipated the existence of preferred or privileged frames, known as inertial frames, as implied by the dynamics of Galileo and Newton. The conservative element in the model was of course the assumption of circular orbits, later corrected by Kepler.

(b) The discrepancy between kinematics (no preferred frames) and dynamics (3-parameter family of preferred frames, also called *uniform motion equivalence*) in Newton's theory has been noticed and objected to by both Leibniz and Huygens, but by advocating dynamical equivalence of all frames (instead of a new kinematics that would yield the preferred frames of Newton's dynamics) they proved in fact more conservative than Newton: the new kinematics yielding the preferred frames was eventually established by Einstein's SRT. Thus, in this particular point, Newton has anticipated Einstein.

(c) In stressing the devotion of the Renaissance scientists to observation, experimentation, and analysis their power for theorizing is often overlooked or underrated. A typical example is Hooke who is well known for his simple law in elasticity theory but not for his synthetic theory of physical optics which even explains the emission of light as due to oscillations in the emitter.

2.2. *Maxwell's Theory: A Change in Ontology?*

Maxwell's theory has been hailed by no less a man than Einstein (1946) as a 'revolutionary transition from forces at a distance to fields as fundamental variables'. I don't think that this is quite correct. On the one hand, conservative forces can always be derived from a potential function over spacetime, on the other hand the same is true of

Maxwell's field quantities. Furthermore, plenum theories of light propagation without action at a distance have been the norm rather than
the exception. The early authors such as Hooke were rather non-
committal as to the nature of the plenum and it seems to be unwar-
ranted to interpret them in the spirit of the later mechanical reduction-
ists working on a mechanical aether theory. I think the revolutionary
character of Maxwell's theory is of a more subtle kind and resides in
the introduction of a new fundamental constant 'c' which turned out to
possess a double meaning: velocity of propagation of electromagnetic
waves in vacuo, and ratio between the so-called electromagnetic and
electrostatic units. The latter meaning demands that c be invariant
under the group of permissible transformations while the first meaning
excludes the Galileo transformation as a permissible transformation.
There are only two ways to solve this dilemma: either we abandon the
equivalence of frames in uniform motion (classical aether theory) or
else we replace the Galileo transformation by a transformation implying the existence of an invariant speed, as suggested by Einstein in
1905. The first solution would establish a true plenum: a single
preferred frame in $E_3 \times T$ could not be explained otherwise. The
second solution, known under the somewhat misleading name 'SRT',
proved to be the physically correct one but is does not imply a plenum
in the original meaning of the word. What it does imply is the existence
of a *single family* of *kinematically* preferred frames (single uniform
motion equivalence), in contrast to all previous physics, but identifiable
with the *dynamically* preferred inertial frames of Newtonian
mechanics.

Of course, the synthesizing power of Maxwell's electrodynamics is
rivaled only by Newton's mechanics. But while the latter has a
somewhat programmatic character, demanding as it does force laws of
a specific kind, Maxwell's theory (supplemented by the Lorentz force
law) is completely free of such programmatic demands. On the other
hand, it is a dualistic theory in the sense that the sources of the field
are charged particles and that the source-free (homogeneous) field
equations do not hold within these particles. Furthermore, all attempts
at remedying this 'defect' (which I do not consider a defect at all) have
proved futile. Thus, if we consider the electromagnetic field as a proper
physical object as we should, Maxwell's theory implies a dualistic

ontology. But this can hardly be called a revolution in ontology: dualistic ontology has a long tradition in physics, apart from the fact that Newton's theóry of gravitation can also be interpreted in a dualistic fashion as mentioned above.

2.3. SRT: A Revolution in Kinematics (Spacetime Structure)

Basically, Einstein's SRT is a revolution in kinematics or spacetime structure: all other changes brought about by this theory spring from this root. The change in spacetime structure consists in the replacement of the direct product $E_3 \times T$ by the pseudo-Euclidean spacetime E_{3+1} and this change is revolutionary in a threefold sense: it is a change in the mathematical substructure, it extends to all physics (with the exception of thermostatics), and it has not been anticipated by any physicist or philosopher (though Poincaré came near to it). Maxwell's theory, or the law of light propagation deduced from it, served as obstetrician, but it is neither sufficient nor necessary for establishing SRT. The basic postulate is the old (so-called) Principle of Relativity demanding the existence of (at least) one uniform motion equivalence, i.e. the existence of a 3-parameter family of global frames such that the transformations between them form a 3-parameter group with the components of the relative velocity as parameters. This implies the existence of an invariant I of dimension (velocity)$^{-2}$ which is either zero, which yields the Galileo transformation, or c^{-2} which yields the (so-called) Lorentz transformation characteristic of E_{3+1}, or $-C^2$ which also yields a pseudo-Euclidean spacetime but with a different global topology, viz., closed space (Strauss, 1972, Chap. 14). Moreover, from the point of view of group theory the special Lorentz group L_x is nothing but the irreducible true representation in (x, t)-space of the algebra characterizing the composition properties of collinear velocities, while the special Galileo group G_x is a reducible representation of that algebra with T as invariant subspace. Thus, every mathematical physicist could have found the Lorentz group if he had studied the composition algebra of collinear constant velocities.

Now a word must be said about the fact that spacetime E_{3+1} (known as Minkowski space) admits one and only one family of preferred frames. On the one hand this is gratifying as it provides a post factum justification of Newton's inertial frames as pointed out above. On the other hand, the existence of a family of preferred frames is of course counter-intuitive to anyone who considers physical space and time as something that can be separated from the properties of physical matter. But people with a wrong philosophy cannot be helped anyhow to understand anything of modern physics.

A particular consequence of SRT kinematics is the fact that the spatial geometry in a non-preferred (non-fundamental) frame is no longer uniquely defined, apart from certain 'remnants' depending on the chosen frame. Thus, even in the relatively simple case of a uniformly rotating frame a common frame time, and hence a uniquely defined spatial geometry, does not exist (Strauss, 1974). This proves that kinematics is more fundamental then geometry – a lesson still to be learned by the majority of philosophers of science, the specialists included.

In all other respects SRT is quite conservative: it preserves Euclidean geometry, though only for the fundamental frames, and it preserves all laws of conservation, apart from that of mass, mass now being part of energy. As to Maxwell's equations, the transformation laws for the field quantities required to ensure covariance under the Lorentz group need not be postulated: they follow from the transformation law for force and the Lorentz force law. (The proof is easy, but I could not find it in any of the familiar text books.)

SRT physics is not a generalization *sensu stricto* of pre-SRT physics but a special form of its dialectic negation. As far as there is generalization, it concerns the concepts rather than the laws. (Cf. Strauss, 1972, Chap. 15.)

In principle, SRT extends to all of physics because c, the limiting velocity, is not only a fundamental constant (like e or the Boltzmann-Planck constant k) but a *universal* constant. But there is of course no point in developing a transformation theory for thermostatic quantities which by definition refer to a privileged frame, viz., the rest frame of the walls between which the thermostatic system is supposed to be enclosed. No wonder that the various authors trying to play this game do not agree on its rules. To be sure, SRT has to be taken into account when – in addition to the *general* laws of thermostatics – *specific* laws are to be used, such as the law connecting energy density with pressure in an ideal gas, i.e. laws that can be derived only from a statistical-mechanical model.

As to gravitation, it is not at all impossible, and indeed rather easy, to generalize Newton's theory in such a way that it becomes Lorentz covariant. But such a theory is just not good enough because gravitation, being a universal property of matter, should be accounted for in a different and more fundamental way, such as in Einstein's GRT. Nevertheless, the Lorentz covariant gravitation theory could serve the useful purpose of comparison with Einstein's theory or any of its many modifications: the differences to the former could then be taken as characteristic of the theory considered. I add this as a further suggestion to the list of meta-physical research programs contained in Strauss (1976).

2.4. GRT – A Revolution in Method?

GRT, i.e. Einstein's chronogeometrical theory of gravitation, has often been presented as example, and indeed the first instance, of a fundamental change in method, to wit: mathematical construction instead of inductive generalization. A certain change in method is indeed undeniable, but the view that this change constitutes a fundamental break with established method is superficial and misleading. True, observational facts requiring a new theory of gravitation were then non-existent or, as Mercury's perihelion shift, not compelling. On the other hand, the theoretical situation established by SRT made Newton's theory untenable. Hence one had to take a new look at the old facts. Now I am not going to repeat the familiar story of how Einstein was led to his new theory – I have analysed some of its aspects elsewhere (Strauss, 1972, Chap. 15). What we are here concerned with is not methodography but methodology of theory construction. One of the 'secrets' (Strauss, 1972, Chap. 18) of theory construction has always been to account for what are considered the most fundamental properties of matter not by the superstructure (equations of motion) but at a deeper level. For Newton, the most fundamental fact was the existence of Galileo's inertial frames in which forceless motion is uniform. Although this follows from his Second Law, he must have been quite aware that this was the very basis of his mechanics and so he formulated it *expressis verbis* as his First Law. A deeper (purely kinematical level, as in SRT) was not available at that time. Similarly, ART starts from the fact that gravitation is a fundamental and universal property of physical matter and not a specific or 'differential' (Reichenbach) force; hence it should be accounted for at the deepest level, i.e. spacetime structure. Thus, the rigid spacetime of SRT had to be replaced by a pliable one that would adjust itself to the distribution and motion of matter. The most conservative generalization of the pseudo-Euclidean spacetime of SRT satisfying this condition is the pseudo-Riemannian spacetime chosen by Einstein, and the superstructure preserving the maximum amount of Newtonian features is indeed given by the Einstein equations of 1916 as shown by many authors. Thus, Einstein's GRT basically does conform to established methodology. Those who hold the opposite opinion either overlook some of the facts just mentioned and/or forget that the facts to be taken into account in theory construction are the more fully represented by existing theories the higher the stage of evolution already attained. If one wishes to stress the novel features of theory construction method thereby implied one should speak of the *growing role of metatheoretical considerations* but not of a revolution in method.

I refrain from considering the other aspects of the change from SRT to GRT as this is hardly any longer a controversial subject among experts.

2.5.   *QM – A Revolution in Logic and/or Ontology?*

Of all changes in fundamental theory that from classical to quantum mechanics is no doubt the most profound one, at least when the mathematical substructures are compared. To start with, there is no point at all in comparing QM with the Newtonian form of CM because there is nothing in the M_{sub} of the former that would in any way correspond to the M_{sub} of the latter. Fortunately, there exist two formulations of CM that can be compared to QM, namely the familiar Hamiltonian and the less familiar Poisson formulation rediscovered by Dirac. Now the most interesting characteristic of these two formulations is the fact that of all known formulations of CM they possess *the widest convariance group*, known as the *group of canonical transformations*. Yet this group has no physical meaning within CM, in particular it does not entail any new laws of conservation: if it did, the

Newtonian formulation would be incomplete. Hence, from the physical point of view these two formulations must appear as highly artificial. However, from the historiological point of view they are instances of what I have called *partial formal anticipations* of a later theory. Indeed, in QM the canonical transformations are a *subgroup of the unitary group*, i.e. the group of rotations in Hilbert space, which is the covariance group of QM. The most important invariants of this group are the angles between two rays; they determine the transition probabilities between the states represented by the rays according to the formula

(1) $\text{prob}_2\,(S_1 \leftrightarrows S_2) = \cos^2\,(r_1, r_2) = \text{Tr}\,\Pi_{r_1}\Pi_{r_2},$

where Π_r is the projector (projection operator) onto the ray r, and Tr is the trace operator. This formula, simple though it is, contains the essence of QM, including the *PI*. It is easily generalized to cases where the rays have to be replaced by n-dimensional subspaces of the Hilbert space:

(2) $\text{prob}_2\,(P_\Pi; P_{\Pi'}) = \text{Tr}\,\Pi\Pi'/\text{Tr}\,\Pi.$

The *P*-symbols are here predicates with the following meaning:

> $P_\Pi(s, t) = $ At time t the state of system s is such
> that the ray representing it lies in the
> subspace onto which Π projects.

The introduction of these predicates is of course the first and decisive step in a formalisation of quantum mechanical language. Moreover, it is an option for the first (Goettingen) Von Neumann (1932, p. 130) and against the second (Princeton) Von Neumann (Birkoff-Von Neumann, 1936). Since the projectors form an atomic partial Boolean algebra, the same is true of our predicates. Inconnectible predicates, i.e. predicates belonging to different Boolean subalgebras, are complementary in the sense of Niels Bohr; they represent properties that can not coexist. Hence the probability relation between any two such properties is a *transition probability* ($prob_2$) and not a *probability of coexistence* of properties ($prob_1$) as considered in the classical theory and its axiomatization by Kolmogorov. As shown elsewhere (Strauss, 1973), a fairly general theory of transition probability can be established without recourse to quantum mechanics. This theory is a genuine generalization of the old (Reichenbach-Kolmogorov) theory to which it reduces in the Boolean subalgebras. The generalization is simply due to *admitting dispositional properties:* if we admit predicates representing dispositional properties it turns out that *classical logic is no longer semantically closed* since it admits the construction of meaningless expressions out of meaningful ones (Strauss, 1973, p. 613). To remedy this defect it is necessary and sufficient to replace classical logic by a logic with partial Boolean algebra. Thus, there is no need to consider any of the other competing 'quantum logics': in the 'Labyrinth of Quantum Logics' (van Fraassen, 1975) they are the blind alleys, even when advocated by respectable men.

Thus, the general theory of transition probability (which is a purely logical theory) is the *logical substructure* of QM in about the same sense in which pseudo-Riemannian geometry is the mathematical substructure of GRT. From this logical substructure the *mathematical substructure* of QM is obtained by choosing the projectors in a Hilbert space as mathematical model of the partial Boolean algebra of predicates. (The other models or representations known so far are less general, but there may exist even more general representations not yet known.)

Now the change from classical logic to a logic with partial Boolean algebra can hardly be called a 'revolution in logic' since it is demanded by classical logic itself if we admit predicates representing dispositional properties and insist on semantic closure. Moreover, it is the most conservative change imaginable; in particular, it preserves semantic 2-valuedness, in contrast to the much advertised Birkhoff-Von Neumann logic which discards 2-valuedness together with the distributive law.

Let us now turn to the ontological aspect.

According to the strategy outlined in Section 1, we have to look for physical objects which can possess the properties represented by the descriptive predicates of the theory. In the case of QM these predicates do not differ from those of CM, apart from the fact that they represent *dispositional* properties. This I would not call a change in ontology but an *advance in ontological dialectics:* a 'dispositional', viz., a *reactive* or *responsive* property implies, in addition to the physical object having that property, also the existence of something to which that object reacts or responds. This, and nothing else, is the ontologic implication of QM. And if you look at the applications of QM you will find that this something is always a macroscopic object describable by classical physics, as insisted by Bohr, and identical with what I have called a *stochastic state changer* (Strauss, 1972, p. 252) to bring out its proper role in quantum physics. Thus, stochastic reaction to external objects is an inbuilt feature of QM although these external objects are not represented explicitly in the formalism of the theory – one source of the many misunderstandings and misrepresentations of QM.

As to the *quantum mechanical concept of 'state'* it follows that it is defined only by the transition probabilities, and since the latter are between states the definition would be circular if none of the states could be defined otherwise. The states that can be defined otherwise are the eigenstates of these so-called observables that correspond to classical variables. Without this correspondence QM would not work. Thus, the semantics of QM implies a certain amount of classical physics – a novel dialectical feature in theory construction. It follows that there is no need, and indeed no place, for a separate 'quantum theory of measurement': a measurement of a quantum mechanical 'observable' is simply the application of a stochastic state changer which compels the system concerned to jump into one of the eigenstates of that observable. (For a more detailed account cf. Strauss, 1972, Chap. 22.)

Among the *conservative features* of QM may be mentioned the existence of a state space and the use of a Hamiltonian for characterising the structure of a physical system and for determining time-dependence.

The *extent* of the quantum theoretical revolution is the same as that of the change from the classical to the Einstein-Minkowski spacetime, i.e. it includes both particle mechanics and field theories (with the probable exception of gravitational theory). Furthermore, QM does not affect the general laws of thermostatics but has to be taken into account when the special features of a thermostatic system have to be considered.

There is one aspect, in which the change from CM to QM resembles the change from CM to ART rather than the change from CM to SRT: the constant 'h' and the gravitational constant appear only in the M^{sup} of QM and ART, respectively, while 'c' appears in both the M_{sub} and the M^{sup} of SRT. I do not think that this signifies anything else but the fact that both 'h' and the gravitational constant are *dynamical* constants while 'c' is a *kinematic* constant. Thus, the consideration of physical dimensions, scorned by so many mathematical physicists, is quite useful even in *meta*-physical research; it should be given more attention by the philosophers of science.

2.5. *QFT - The Neglected Revolution*

Quantum field theory really started in 1900 with Planck's famous law, 25 years before QM was established (cf. Strauss, 1972, Chap. 4, Sections vii–x), but it could not develop any further until QM could be used as a guide of how to quantize fields. I recall the following points:

(i) The theoretical foundation invented by Planck contained *two innovations*, the well-known *quantization of energy* and the use of a *novel kind of statistics* that came to be known as Bose-Einstein statistics – one of the many misnomers in metascience. Oddly enough, the quantization of energy attracted much more attention than the novel kind of statistics although it is implicit in Wien's radiation law just as the so-called Planck constant h; in fact, it was by comparing Wien's radiation law with the measurements of Kurlbaum and Paschen that Planck had found the value $h = 6.885 \ 10^{-27}$ erg sec. Odd enough, too, that Planck's novel kind of statistics was strongly criticized (even by Einstein) instead of being considered as a plausible innovation, while the obvious inconsistencies in Einstein's 1905 paper founding SRT (Stiegler, 1974) were generally overlooked. The Planck statistics – to give it its proper name – is equivalent to the assumption that the elements concerned have no individual traits by which they could be distinguished so that a conceptual permutation of the elements corresponds to no real change in the situation and hence does not yield a new 'complexion' (i.e. microstate), contrary to the old Boltzmann statistics. The *ontologic implication* of this is rather obvious: the *elements concerned* (here: the energy quanta of the electromagnetic field) *do not possess a location in space or a worldline in spacetime.*

(ii) Planck's radiation law, if correct, implies that the correct theory of electromagnetism must be some kind of *synthesis of the two classical theories of light* which were in contest with one another for centuries. This follows from Einstein's (1909) discovery that the thermostatistical fluctuations of energy in a partial volume, implied by the Planck law, consist of two parts, one corresponding to classical wave theory and the other to classical particle (ideal gas) theory.

Today, QFT is of course a well-established and indeed the most advanced part of theoretical physics since it combines the basic requirements of both SRT and quantum theory. However, it is not free of internal difficulties which are widely analysed but not understood as consequences of postulates that could be given up without destroying the achievements of the theory. Hence we shall concentrate on the achievements of QFT which can be expected to survive any future change in theory.

The basic achievement of QFT is of course the dialectic synthesis of the two classical ontologies (fields and particles) already implicit in Planck's radiation law. I call this synthesis 'dialectic' because it results from a dialectic negation of both classical fields and classical particles. Needless to say that the operative role in this synthesis is played by what Bohr has called 'the complementary mode of description' and what we now call 'logic with partial Boolean algebra'.

The next most important achievement of QFT is the explanation (first given by Pauli) of the empirical relation between spin value and permutational parity and, hence, between spin value and the kind of 'statistics' (Bose-Einstein or Fermi-Dirac) which cannot be explained within QM. Since Fermi-Dirac statistics implies Pauli's exclusion principle we can now understand why fermions (e.g. electrons, protons, neutrons) behave more nearly like classical particles than bosons (e.g. photons and – if they exist – gravitons) of which an unlimited number can (and usually does) occupy the same state, which corresponds to a continuously variable field strength. Thus, *the two classical ontologies can now be considered as partial and approximate anticipations of QFT ontology.*

Further, since particles are conceived as field quanta in QFT the number of particles of a given kind and state is not a constant but an 'observable' having eigenvalues $0, 1, 2, \cdots n, \cdots$ for tensor fields (boson fields) and $0, 1$ for spinor fields (fermion fields). Hence mutual transformation of particles of different kind becomes possible, subject to the conservation laws.

Besides the old conservation laws following from the symmetries of Minkowski spacetime (or Lorentz covariance) there exist a number of *additional conservation laws* which express so-called *internal symmetries* related to so-called internal degrees of freedom. This is the borderline where QFT goes over into the theory of elementary particles, which is still in the making.

In spite of its far reaching implications, the QFT of tensor (boson) fields is a rather conservative theory in comparison to SRT and QM; in particular, it satisfies Bohr's principle of correspondence as alluded to above. Indeed, the quantization of boson fields is little more than the application of the quantization rules known from QM. By way of contrast, the quantization of spinor fields is effected by using anticommutators instead of commutators. Hence there is no simple correspondence linking quantized spinor fields to either classical fields or classical particles. (A detailed study of the pertaining intertheory relations is still missing.)

This concludes my survey of the great revolutions in physics. Sketchy though it is it will have demonstrated that the evolution of physical theory is a complicated dialectical process in which both innovation and conservation play equally important roles. Hence, to speak of a 'paradigm change' is both superficial and misleading: where the explications of that model are true (in the sense of a zeroth approximation) they are trivial, and where they are non-trivial they are wrong. Even if they were right the model would be useless since it does not contain any heuristic guide how to solve future crises in physics: laws governing the evolution of physics (Strauss, 1972, Chap. 2) have no place in that model and are implicitly denied; instead, we are offered the inventors of new paradigms as veritable *dei ex machina* who establish themselves by some kind of persuasion, only to be overthrown by a new *deus* with even more persuasive power. Besides, what we need at present is not a general model of theory change (which is bound to be poor in content if it is general enough) but more detailed studies of intertheory relations and evolutionary laws to be abstracted from them.

3. WHAT ABOUT FUTURE REVOLUTIONS IN PHYSICS?

Only a few remarks can be made here on a subject that does not yet exist and about which all conjectures may appear equally unfounded. Still, the situation is not quite as hopeless as it may appear. As a guide to the future we can use the well-established evolutionary laws (Strauss, 1972, Chap. 2) in connection with an analysis of present – internal and external – difficulties. The trouble is that – so far – there are hardly any external difficulties while the internal difficulties are either overcome (e.g. renormalization in QFT) or inconclusive, viz., without definite heuristic value. In spite of this, or perhaps because of this, new theories, theory fragments or approaches appear year after year, and sometimes month after month. A large part of them is due to philosophic dissatisfaction with the dialectic features of modern physics and hence retrogressive rather than progressive; they usually have a short life and are soon forgotten. Among the remaining attempts the most interesting and progressive group is characterised by the endeavour to construct *stronger* and/or *more adequate* theories. A theory T' is *stronger* than T if it explains more than does T, and it is *more adequate* if it accounts for some fundamental property of matter in a more fundamental way.

A fundamental property of matter is the existence of *two essentially different kinds of particle-fields:* tensor fields yielding Bose particles and spinor fields yielding Fermi particles. This is properly accounted for by QFT, but not by GRT which knows only tensor fields. Hence an *adequate* representation of the action of the gravitational (metrical) field on spinor fields or their particles is not possible within the Einstein-Riemannian spacetime but requires a *bimetrical spacetime*, to wit, a spacetime carrying both a pseudo-Riemannian and a Minkowski metric. This is the basic idea underlying Treder's (1967a,b, 1969) *tetrad theory of gravitation.*

Properties of matter that ought to be explained by a good theory of

M. STRAUSS

elementary particles are (among others): (i) the *mass spectrum*, i.e., the empirical connection between the rest mass of a particle and its (internal and external) symmetries, and (ii) the empirical values of the coupling constants which determine the interaction between the various particle-fields and hence the mean life time of the many unstable particles. The most determined attack on this problem is as yet Heisenberg's 'unified' (primeval) spinor field theory of fundamental particles. The primeval spinor field satisfies a *non-linear* differential equation and hence contains a *new fundamental constant* having the dimension of *length* and has as many symmetries as are required to account for the 'strong' conservation laws. The primeval field itself (or its particles) cannot be observed – it corresponds to the $\alpha\pi\epsilon\iota\rho\text{o}\nu$ in the natural philosophy of Anaximenes.

I have selected these two theories because I think they are typical of what we have to expect in the future. On the one hand, they conform to the evolutionary laws (Strauss, 1972, Chap. 2) and the strategy based on them, on the other hand their relation to the preceding theories is of a rather different kind from the intertheory relations known and studied so far; in particular, there is no correspondence relation in the old familiar sense.

Will the advance in fundamental theory ever come to an end? I do not think so, and this for two reasons. First, we must expect that more powerful and/or more refined means of experimentation and observation will reveal new properties or even new kinds of matter. Second, a new theory usually creates new problems, even if it solves the old ones. (Thus, Heisenberg's 'unified' theory of fundamental particles requires a Hilbert space with *indefinite* metric, which creates problems of physical interpretation and consistency.) There will always be place for more adequate, stronger, or better organized theories.

Berlin, D.D.R.

BIBLIOGRAPHY

Czelakowski, J.: 1974/75, 'Logics Based on Partial Boolean Algebras (1) and (2)', *Studia Logica* 33, 371–396 and 34, 69–86.

Einstein, A.: 1949, 'Autobiographical Notes', in P. A. Schlipp (ed.), *Albert Einstein: Philosopher-Scientist*, The Library of Living Philosophers, Inc., Evanston.

Furmanova, O. V.: 1973, 'Vnutrennjaja logica dviženija nauki i prognozirovanie es razvitija', *Naučnoc upravlenie obsčestrom* **7**, 120–128.

Heisenberg, W.: 1974, 'The Unified Field Theory of Elementary Particles: Some Recent Advances', *Naturwissenschaften* **61**, 1–5.

Hooker, C. A. (ed.): 1973, *Contemporary Research in the Foundations and Philosophy of Quantum Theory*, D. Reidel Publ. Co., Dordrecht and Boston.

Hooker, C. A. (ed.): 1975, *The Logico-Algebraic Approach to Quantum Mechanics*, D. Reidel Publ. Co., Dordrecht and Boston.

Koertge, N.: 1973, 'Theory Change in Science', in G. Pearce and P. Maynard (eds.), *Conceptual Change*, D. Reidel Publ. Co., Dordrecht and Boston.

Kuhn, Th. S.: 1962, *The Structure of Scientific Revolutions*, The University of Chicago Press, Chicago.

Mostepanenko, A. M.: 1974, *Prostranstvo i vreja v makro-, mega- i mikromire*. Moskva: Isd. pol. lit.

Stiegler, K.: 1974, 'On Errors and Inconsistencies Contained in Einstein's 1905 Paper 'Zur Elektrodynamik bewegter Koerper'', in *Proceedings of the XIII. International Congress for the History of Science*, Section VI, Editions 'Naouka', Moskwa.

Strauss, M.: 1972, *Modern Physics and Its Philosophy – Selected Papers in the Logic, History, and Philosophy of Science*, D. Reidel Publ. Co., Dordrecht.

Strauss, M.: 1962, 'Verallgemeinerung des Planckschen Strahlungsgesetzes in h-c-l-Theorien', *Zeitschrift für Naturforschung* **17a**, 827–847.

Strauss, M.: 1973, 'Two Concepts of Probability in Physics', in P. Suppes *et al.* (eds.), *Logic, Methodology and Philosophy of Science IV* (Proceedings of the 4th International Congress for Logic, Methodology and Philosophy of Science, Bucharest, 1971), North-Holland Publ. Co., Amsterdam and London.

Strauss, M.: 1974, 'Rotating Frames in Special Relativity', *International Journal of Theoretical Physics* **11** (1974) 107–123.

Strauss, M.: 1976, 'Meta-Physical Research: Reflections and Suggestions', to appear in C. A. Hooker (ed.), *Abstract Presentation in Mathematical Physics* (working title), The University of Western Ontario Series in Philosophy of Science, D. Reidel Publ. Co., Dordrecht and Boston.

Suppes, P. (ed.): 1973, *Space, Time and Geometry*. D. Reidel Publ. Co., Dordrecht and Boston.

Treder, H.-J.: 1967, 'Das makroskopische Gravitationsfeld in der einheitlichen Quantenfeldtheorie', *Annalen der Physik* (7) **20**, 194–206.

Treder, H.-J.: 1967b, 'Das Gravitationsfeld in der einheitlichen Quantenfeldtheorie', *Monatsberichte der Deutschen Akademie der Wissenschaften zu Berlin* **9**, 283–292.

Treder, H.-J.: 1969, 'Heuristische Begruendung einer Tetradentheorie des Gravitationsfeldes', *Annalen der Physik* (7) **22**, 201–204.

Treder, H. -J.: 1974, *Philosophische Probleme des physikalischen Raumes*, Akademie-Verlag, Berlin.

Van Fraassen, B. C.: 1975, 'The Labyrinth of Quantum Logics', in C. A. Hooker (ed.), *The Logico-Algebraic Approach to Quantum Mechanics* I, D. Reidel Publ. Co., Dordrecht and Boston.

MICHAEL FREDE

THE ORIGINS OF TRADITIONAL GRAMMAR

By 'traditional grammar' I mean the kind of grammatical system set
out in and presupposed by standard modern grammars of Greek and
Latin like Kühner-Gerth or Kühner-Stegmann. Since grammars of this
kind traditionally have been followed quite closely by grammarians of
other languages one may speak traditional grammar quite generally.
Grammars of this type consist of three parts: a phonology, dealing
among other things with the sounds of the language, a morphology,
dealing with word-formation and -inflection, and finally a syntax in
which we are told which combinations of words constitute a phrase or
a sentence. Moreover, such grammars are characterized by a certain
set of concepts, especially the so-called grammatical categories, that is,
notions of various parts of speech like that of a noun or a verb, and
notions of various features of these parts of speech which traditionally
are called 'accidents' or 'secondary categories'. Examples of such
accidents are gender, number, case, mood, tense.

The only ancient text which more or less fits this characterization is
Priscian's Institutiones, written in the early 6th century A.D. But we
do know that earlier ancient grammarians like Apollonius Dyscolus in
the 2nd century A.D. covered in their writings the whole of traditional
grammar, and we can see from his extant writings on various parts of
speech and on syntax that by his time the concepts of traditional
grammar were already fairly well established. Hence we have to look
for the origins of traditional grammar in antiquity.

Some of the concepts of traditional grammar can in fact be traced
back to the 5th century B.C. Protagoras distinguished the three
genders (Aristotle *Rhet.* 1407b6; cf. *Soph. El.*173b17), and during the
following centuries the familiar concepts seem to make their appearance one after the other. Hence historians of ancient grammar like
Steinthal tend to treat their subject as if it had a more or less
continuous history. If one treats of the subject in this way, though, one

Butts and Hintikka (eds.), Historical and Philosophical Dimensions of Logic, Methodology and Philosophy of Science,
51–79.
Copyright © 1977 by D. Reidel Publishing Company, Dordrecht-Holland. All Rights Reserved.

runs the risk of overlooking the motives and guiding principles followed by those who first tried to write something like a systematical grammar as opposed to those who just occasionally would care to note a point of grammar. Hence, when I talk of 'the origins of traditional grammar' I am trying to talk about the first attempts to write something like a traditional grammar.

Unfortunately the origins of traditional grammar in this sense are very obscure, mainly because all the texts of the formative period, with one exception, have been lost. The exception is Dionysius Thrax' Τέχνη γραμματική, and this text poses such problems that even its authenticity has been doubted (cf. Di Benedetto, Dionisio Trace e la Techne a lui attribuita, Annali, Scuola Norm. Sup., Pisa, v. 27, 1958, p. 169 ff.), though nowadays the text is generally accepted as genuine. But the origins of grammar are also very much obscured by an ambiguity in the ancient terms for grammar or the grammarian. Grammar may just be the modest art of reading and writing, sometimes also called 'small' or 'lower grammar' (Scholia Vaticana in Dionysium Th. 114, 23 ff. Hilgard; cf. ἀτελεστέρα Philo De congr. erud. causa §148) or 'γραμματιστική' (Sextus Emp., A.M. I,44; Philo ibid.). In Hellenistic and later times 'grammarian' primarily refers to those Alexandrian scholars and their followers, including the schoolmasters, who are concerned with the restitution, the proper reading, the explanation and interpretation of the classical texts, and their literary criticism. This art was also called 'great' or 'higher grammar' (Schol. Vat. 114, 27ff), 'perfect' or 'complete grammar' (e.g. ἐντελής S.E. A.M. I,44; τέλειος I,46; τέλεια I,76; τελειοτέρα Philo l.c.). High grammar is basically what we call 'philology'. And it was in fact one of these Alexandrian scholars, Eratosthenes, who first wanted to be referred to as a 'philologos' (Sueton. De gramm. 10, 4). Finally, incorporated as a part of this grammar, we find a discipline called 'the technical part of grammar' (S.E. A.M. I,91; 96) or 'methodical grammar' (Quintilian I,9, 1). It is this technical grammar which corresponds to our traditional grammar; and hence what we are concerned with are the origins of this technical grammar.

Since the earliest grammar of this type we know of is the one I referred to above, written towards the end of the 2nd century B.C. by the Alexandrian scholar Dionysius Thrax and often thought to be the first grammar ever to be written in our tradition (cf. Susemihl, *Geschichte der griechischen Literatur in der Alexandrinerzeit*, II, p. 170 and more recently R. H. Robins, *Dionysius Thrax and the Western Grammatical Tradition*, Transactions of the Philological Society, 1957, p. 67) and since technical grammar subsequently seems to fall within the domain of scholars and grammar-school masters it is natural to assume that grammar had its origins in the tradition of Alexandrian scholarship (cf. Robins, 'The development of the word-class system of the European grammatical tradition', *Foundations of Language* 2, 1966, p. 6). But, though it seems safe to assume that grammar owes its independence, as a subject on a par with rhetoric and dialectic, to this tradition as it was received in the schools, a look at the facts makes it difficult to believe that it owes its origin to Alexandrian philology.

Dionysius Thrax' *Techne* starts out with a definition of grammar: "Grammar is knowledge by experience of what is said for the most part by poets and prose-writers". He goes on to specify six parts of grammar: (1) the skilful reading of texts, (2) the explanation of poetical tropes, (3) the explanation of peculiar words and realia, (4) etymology, (5) analogy, i.e. the explanation of declined and, perhaps, conjugated word-forms, (6) literary criticism, the finest part of grammar, as Dionysius adds. It may be easier to understand this list if we assume that it is a development of an earlier list of four parts referred to by the scholiasts (e.g. p. 12, 3ff.) and Varro (ap. Diomedem, Grammatici Latini I, 426 Keil), though they do not agree on the order: (1) the correction of the text, (2) its reading, (3) its explanation, and (4) its criticism, i.e. its evaluation. It is obvious that these four parts correspond to the topics somebody in philology would go through in lecturing and commenting on literary texts. Hence Dionysius in the introduction seems to think of Alexandrian philology when he promises us a treatise on grammar. For the next four paragraphs our expectations are fulfilled, for he talks of reading, accents and punctuation (and has three lines on rhapsody). But then, without any warning or explanation, he starts with a treatise of technical grammar covering letters, the corresponding sounds, syllables, and then the parts of speech. Not the slightest explanation is given as to how this technical grammar fits the notion of grammar as it is defined in the first sentence or how it is related to the six parts of grammar referred to. The best explanation

for this abrupt juxtaposition seems to be the following: at the time
Dionysius wrote his Techne there was no established systematical
connection between philology and technical grammar such that a few
sentences, to keep in style with the rest of the text, would have sufficed
to explain what role grammar plays in philology.

It is, of course, possible that philology and technical grammar had
been pursued in the Alexandrian school alongside each other for some
time and that only now Dionysius tried to incorporate the study of
technical grammar into some philological curriculum. After all, given
the nature of the philological enterprise, one hardly gets along without
the use of grammatical notions. But if we look at the testimonies
concerning the work of Dionysius' Alexandrian predecessors there is
no hint that any of them wrote on technical grammar. It is only for
Dionysius' successors that treatises on technical grammar are attested,
e.g. one on the parts of speech for his student Tyrannio (Suida s.v.).
Claims as to the achievements of Dionysius' Alexandrian predecessors
in this area are usually supported by reference to the following facts:

(1) Aristarchus, Dionysius' teacher, is said to have distinguished
the canonical eight parts of speech (Quint. I,4,20),
(2) Aristarchus and his predecessor, Aristophanes of Byzantium,
tried to set up rules of declension (Charisius, p. 149, 26ff. Barwick).

And it is, in fact, in this connection that Pfeiffer (*History of Classical
Scholarship*, p. 203) says: "The term 'grammar', so far consciously
avoided, can now indeed be used; we can see that as part of scholarship in general a separate discipline was being built up which reached
its height in the second generation after Aristophanes, in the τέχνη
γραμματική of Dionysius Thrax, the pupil of Aristarchus". In support
of this Pfeiffer refers to Sextus Empiricus who says (*A.M.* I,44) that
'grammar' more specifically is called that perfect discipline worked out
by the followers of Crates Mallotes, Aristophanes, and Aristarchus.
But it should be clear from what was said in the beginning about the
ambiguity of the term 'grammar' that this statement may refer to
philology quite generally; and that it actually does so is brought out by
the etymology of 'grammar' offered by Sextus a few paragraphs later
(I,47f.).

This, then, leaves us with the two facts referred to as the main evidence for Alexandrian contributions to technical grammar prior to Dionysius. So let us consider these in turn. According to Quintilian (I,4,20) Aristarchus assumed eight parts of speech. As we can see from the context, Quintilian refers to the eight parts we also find in Dionysius Thrax and which then came to be commonly accepted for Greek grammar (for Latin interjections were added as a separate class, but the number 8 was preserved, since Latin does not have the definite article; cf. e.g. Quint. I,4,19). It is very difficult to evaluate this testimony. Since Aristarchus does not seem to have written on the parts of speech, the testimony may rest on no more than the observation that Aristarchus in his writings in practice distinguishes eight classes of words which exactly correspond to the eight parts of speech. But from this it would by no means follow that Aristarchus distinguished eight parts of speech. For one might talk of proper names and common nouns, or common nouns and adjectives, and it still might be left open whether one considers these as one or two parts of speech (consider, e.g. the question whether one should say on the basis of *De congr. erud. causa* §149 that Philo distinguished proper name and common noun as two parts of speech). It also may be pointed out that Quintilian does not suggest that it was a major achievement of Aristarchus to establish this canon of eight parts of speech whereas Aristotle, as Quintilian believes (I,4,18), had only recognized three. For he says that the number of parts of speech was slowly raised from Aristotle's times by the philosophers and in particular by the Stoics (I,4,19), and Aristarchus is only referred to as an authority among those who accepted the list of eight parts which later was to become canonical. There is no suggestion that the list was accepted because it was thought that Aristarchus had cleared the matter up. Even his best students, Dionysius and Apollodorus, did not agree, at least at times (Apol. Dysc. *De pron.* 5, 18–19; Schol. Vat. in D.Th. 160, 26–28), nor did Didymus (*Prisc. Inst.* XI,1,1, p. 548, 7ff.; for the whole matter cf. also, e.g. Schol. in D.Th. 58, 21; Quint. I,4,20; Dionys. Halic. *De comp. verb.* 2; Apol. Dysc. *De coni.* 214, 25–26; Ps. Herodian *De soloecismo*, Nauck Lexicon Vindob. p. 295; Papyrus Yale V c.I lines 3 and 10).

Of more substance is the other testimony. According to Charisius (p. 149, 26ff.) Aristophanes and Aristarchus were concerned with the analogy of declinable words. Similar words should be expected to have a similar declension. The question just is with reference to which features we determine the relevant similarity of words. Aristophanes specifies five such features: gender, case, word-ending, number of syllables, and accent; to which Aristarchus added a sixth: figure, i.e. whether a word is simple or composite. Words being the same with reference to these features should have the same declension. Following this approach later grammarians added further features and tried to formulate rules of declension in terms of these.

So there seems to be a concern with declension, out of which an interest in inflection in general would grow naturally. Later inflection was dealt with in treatises on Hellenismos (or Latinitas), i.e. treatises trying to specify what is to count as proper Greek (or Latin). In these treatises analogy and etymology were regarded as the major criteria to judge the acceptibility of an expression. Hence it is tempting to assume that when, as we saw above, Dionysius Thrax in addition to the four regular parts of grammar also lists analogy and etymology he is taking into account a traditional concern of the Alexandrian school at least for inflection. But it is significant that though Dionysius may refer to such studies in his introductory characterization of grammar, he himself in the treatise does not deal with declension. Hence it would seem that the evidence usually relied on for the claim that technical grammar is of Alexandrian origin is rather insufficient.

One could, of course, proceed to list as further evidence grammatical concepts or assumptions Aristarchus or his Alexandrian predecessors used. What such a survey would show is that Dionysius' *Techne* cannot be regarded as an elaboration, let alone as a codification of the grammatical system presupposed by the Alexandrians.

So, to save the assumption of an Alexandrian origin of this kind of technical grammar, one would have to assume that it was developed by Dionysius Thrax himself. But a look at his treatise shows that this hypothesis is extremely implausible. It has the character of a very elementary handbook in which a fairly rich system for didactic purposes has been reduced to basic definitions, classifications and examples of the classes referred to. In fact, it has been discussed as a paradigm of such an elementary handbook in antiquity by Fuhrmann in his "Das systematische Lehrbuch". And we may assume that the

Techne owed its later success to the radical simplification with which it treated its complex material. It is definitely not the kind of text that would be written to introduce a new discipline of such interest and importance.

Now, if one looks for the origin of grammar outside the Alexandrian tradition of philology one has to turn to the philosophers who ever since the 5th century had taken an interest in language. In fact there are traces of Peripatetic influence in ancient grammar. But I think that it can be shown that most of these traces are either due to the influence of the Aristotelians on the early Stoics or are of much later origin when some form of Peripatetic logic became part of what was accepted by any educated person (i.e. at least from the late 2nd century onwards). One of the reasons, it seems, why the medieval speculative grammarians tried to reconstruct grammar was that traditional grammar did not fit their views on language shaped by Aristotle's Categories and De interpretatione.

So we have to turn to the Stoics. It has been argued by Pohlenz (cf. his "Die Begründung der abendländischen Sprachlehre durch die Stoa", *Nachr. Ges. d.W. Göttingen* 3, 1938, p. 151ff.) and in much detail by Barwick that traditional grammar is of Stoic origin. Pohlenz argued in particular that the case-system and the verbal tense-system are basically Stoic. This conclusion seems to be generally accepted, and hence all authors admit some Stoic influence. But there seems to be a general feeling of uncertainty concerning the extent of this influence. Some authors tend to minimize it. Robins, e.g. says "there seems to be little to substantiate the alleged Stoic influences" (*Dionysius Thrax* ··· p. 76n4), in support of which he refers to M. Schmidt's famous paper on Dionysius in Philologus, 1853. Other authors show a significant tendency to waver on this point from one page to another. This scepticism and the uncertainty to a good extent seem to be due to the fact that we still know so little about the details of Stoic grammar or even its scope. Hence the Stoics' contribution to the subject might still seem to be insignificant and incidental.

The difficulties begin with the fact that among the many parts of philosophy recognized by the Stoics there is none which is called 'grammar' nor is there one which could at least be identified as grammar. Those who claim Stoic origin for grammar point to the τέχνη περὶ φωνῆς by Diogenes of Babylon as the probable source for handbooks of grammar like that by Dionysius Thrax. But though this may be right it would be a mistake to think that Stoic grammar is to be identified with that part of Stoic dialectic which deals with human or Greek utterances or any part thereof.

The Stoics standardly divided philosophy into logic, physics, and ethics. One may say that logic for them is the doctrine of what somebody says who is guided by reason. Traditionally logic was divided into rhetoric and dialectic. But dialectic came to be concerned with rational (Greek) utterances quite generally, at least insofar as their study did not fall into the province of rhetoric. Hence the general study of language would belong to dialectic. Dialectic in turn was divided into two parts, one concerned with the utterances, the other with what was said by uttering these sounds (or writing these letters). To use Stoic language, one part of dialectic is concerned with what is signified or what is said, the other with what is signifying or with voice. It seems to me, then, that it is a mistake to think that Stoic grammar could be identified with that part of dialectic which deals with signs or with a part thereof. For a large number of the traditional notions of grammar like those of a case, casus rectus and obliquus, active and passive, past, present and future, singular and plural, construction or syntax, and many others seem to have been introduced by the Stoics in the part of dialectic concerned with what is signified, and their application to utterances appears to be secondary. Hence let us start with a consideration of the relevant parts of the Stoic doctrine of what is said or signified.

According to Stoic doctrine if one says something two kinds of items are involved. There is (1) the utterance one produces and (2) that which one is saying in producing this utterance. An item of the first kind is a sound of a certain sort, according to the Stoics air modulated or articulated in a certain way. An item of the second kind, on the other hand, is supposed to be immaterial; it is called a 'lekton' by the Stoics, i.e. something which is or can be said. The Stoic doctrine of lekta is a notoriously difficult subject-matter. Lekta in Stoic philosophy seem to have two functions, of which the first tends to be overlooked, particularly by those who are interested in Stoic logic and their philosophy of language. Since this function is of some importance for one point in our investigation I will have to consider it at least briefly.

The first philosopher who is attested to have used the term 'lekton' is Cleanthes: according to Clement of Alexandria (Strom. VIII, 9, p. 96, 23ff.) some (i.e. the Stoics) say that causes are causes of predicates or "as some put it, of lekta, for Cleanthes and Archedemus call these predicates lekta". We know from many sources that for the Stoics both causes and what is affected by these causes are bodies, but that which is effected or brought about by these causes is something immaterial. For what is brought about is that something is so-and-so, e.g. cold, but that something is cold is not a body, though it is a body which is cold. This assumption created a certain difficulty for Stoic ontology. For it was claimed by them, strikingly along the lines of the Earth-born giants in Plato's *Sophist*, that everything which is real, including qualities, is corporeal and that it is the test for the reality of something whether it can act upon or be affected by something. But on the other hand the Stoics did need in their physics certain incorporeal items anyway, namely place, time, and the void. And hence, to cover these and the lekta brought about by causes, they felt forced to introduce the notion of a something (τι) which may subsist though it does not exist or is not real (S.E. *A.M.* X, 218). Behind this recognition of lekta at least under the limited status of a subsistent something lies, I think, a criticism of Aristotle. Aristotle accepts entities like a man or health, also such compound entities as a healthy man. But there seems to be no place in Aristotle's ontology for such items as being healthy or being pink or the fact that Socrates is healthy. It is the colour pink which for Aristotle and the Stoics is a quality and not an item such as being pink. Hence the Stoics think that if Socrates is healthy there are at least three items involved, namely the two recognized by Aristotle, Socrates and the quality health, but then also a third one, being healthy or that *x* is healthy. It is this kind of item which, we are told, they call a 'predicate' or a 'lekton'. As opposed to the first two items it is incorporeal.

This notion of a lekton and in particular of a predicate occurs not infrequently, but primarily in two contexts: in explications of the notion of a cause and in remarks about the proper objects of choice, desire, wishes and the like (cf. Stobaeus *Ecl.* II p. 98, 5W.).

More important, though, seems to be the second function lekta serve. Lekta are also something like the meanings of sentences in the following sense: as we saw above, the Stoics assumed that when we say something two kinds of items are involved, the utterance and the lekton. Since it is also assumed that whenever we say something we either make a claim, ask a question, give an oath, address or invoke somebody, give an order, pray or ask for something or do something else of this kind which involves producing an utterance, the Stoics divided the lekta accordingly into claims or propositions, orders, questions, oaths etc. (S.E. *A.M.* VIII 70; D.L. VII, 65–68). There was some interest in non-propositional lekta. Chrysippus devoted several writings to questions and one in two books to orders (D.L. VII, 191). There are some interesting remarks concerning the relation between oaths and truth (cf. Simplicius *In cat.* 406,23; Stob. *Flor.* 28,14; 28,15). And the badly damaged papyrus with a book of Chrysippus' *Logical Investigations* shows at least that Chrysippus considered utterances like "walk, since it is day" (col. 11n, *Stoicorum vet. fragm.* II p. 107, 36–37 von Arnim), "walk, or if you don't do that, sit down" (12n), "either walk or sit down" (col. 13n). But most of the attention was given to propositions, and hence in the following I will restrict myself to these and the corresponding sentences, except to note that the traditional terms for two of the moods of the verb, 'optative' and 'imperative', seem to be derived from the names of two kinds of such non-propositional lekta, prayers and orders,[1] and similarly the term 'vocative' seems to be derived from the name of another kind, the address.[2] What happens in these cases is something which we will observe again and again: distinctions are made on the level of the lekta which are reflected by certain features on the level of the expressions which features then get named after the corresponding features of the lekta. A standard way to pray for something or to express a wish is to use a sentence with the verb in the optative, to give an order one standardly uses the imperative, and if one addresses somebody one uses the vocative. To say this and to assume such a correspondence is not to commit oneself to the view that to use the optative is to express a wish or that the only way to express a wish is to use an optative.

Parts of the Stoic doctrine of propositions are of interest for our purposes because the Stoics treated even simple propositions as in a way complex (cf. e.g. S.E. *A.M.* VIII, 79; 83; 94). Though, as is well known, Stoic logic is basically propositional the Stoics made considerable efforts to analyze propositions, developed a doctrine of elements of propositions (and lekta in general) and something like a syntax to specify which of these elements would go together in which way to form a lekton. There is every reason to believe that our use of the term 'syntax' is derived from their use for the syntax of lekta (cf. D.L. VII, 192; 193; also compare the corresponding use of συντάττειν, e.g. in D. L. VII, 58; 59; 64 four times; 72 twice). I will return to this point later.

We have some information concerning the syntax of propositions. These lekta are talked of as if they were generated (cf. D.L. VII, 64; Suida s.v. ἀξίωμα p. 255,2 Adler) or put together from the elements. Hence the expressions 'constructio' and 'syntaxis' which literally mean 'putting together'. It seems that this construction was supposed to take place in a certain order. We start with the construction of the predicate we have encountered above. But the predicate is essentially incomplete, in fact, it is called an incomplete lekton (D.L. VII, 64). Hence we proceed to the construction of a simple proposition. From there we may proceed to the construction of various kinds of negative simple propositions (but cf. D.L. VII, 70) and then to the construction of non-simple propositions by combination of the propositions constructed so far.

Let us start with a consideration of the elements of propositions. There is, first of all, the predicate. Since we have talked above about the construction of predicates it is clear that either the term 'predicate' does not refer to a kind of element since predicates themselves may be complex or that the term is ambiguous. It seems that Stoic terminology in this respect is not uniform, but there is at least one terminology according to which the term is ambiguous. For in D.L. VII, 64 it is said that those predicates are direct which, combined with an oblique case, will form a predicate (πρὸς κατηγορήματος γένεσιν). The language here strongly suggests that there is a sense in which the predicate taken without the oblique case is not yet a predicate. And so we do have predicates in the sense of parts of propositions which satisfy certain conditions such that from them one may proceed to the construction of a proposition. These predicates may be simple or complex. In the latter case they themselves would have to be constructed before one could proceed to the construction of a proposition. But then there are, secondly, predicates in the sense of elements of propositions. It is these that we are concerned with right now. Let us call them elementary predicates as opposed to syntactical predicates.

These elementary predicates are what corresponds to the verb in the corresponding sentences. There is a very rich and complex tradition concerning various kinds of elementary predicates. The Stoics distinguished, e.g. between direct predicates, which correspond to active or active transitive verbs, and oblique predicates, which correspond to middle and passive verbs. A distinction between middle and passive then may have been drawn within the class of oblique predicates (cf. D.L. VII, 191; Chrysippus, Q.L., *SVF* II, p. 97, 37–99, 36 von Arnim; D.L. VII, 64). Predicates were also distinguished as to number (Chrysippus, Q.L. *SVF* II, p. 99, 39) and tense. The tense distinction stressed aspect which later tended to be neglected by Greek grammarians, partly due to a change in the language itself.

Another important kind of element of the proposition which like the elementary predicate perhaps is needed for the construction of any proposition (but, again, cf. D.L. VII, 70) is the so-called case (πτῶσις). Several kinds of cases are distinguished. A case is supposed to come either as a casus rectus or a casus obliquus (D.L. VII, 64–65), and among the casus obliqui the Stoics distinguished between the genetive, the dative, and the accusative. It is obvious that our case-system and the corresponding terminology are derived from this.

Among the many problems concerning the doctrine of cases I would like to comment on two closely related ones. The first was already raised in antiquity: why is the nominative a case? The Peripatetics claimed that it was only called a case by courtesy (καταχρηστικῶς) since it was not itself a case of anything in any sense. And this explanation seems to have remained the standard one. But it is almost certainly wrong. For the Stoics insisted that the nominative is a case straightforwardly (Ammonius *In de int.* 43, 4–5; Stephanus *In de int.* 10, 22–26). Hence in late antiquity and in Byzantine times there was a considerable amount of speculation as to what the nominative might be thought to be the case of. Some of this speculation is quite interesting. It leads, e.g. to the abstraction of a word as opposed to its inflected realization. But as far as our problem is concerned all this speculation seems to be off the mark.

The correct solution seems to be suggested by a solution of the second problem. This problem arises in the following way: it is clear that one class of cases corresponds to nouns. In the sentence "Socrates is wise" 'is wise' would correspond to a complex syntactical predicate and 'Socrates' to a case in the underlying proposition. Now predicates are incomplete lekta and as such incorporeal items. Hence one should suppose that cases, being elements of lekta too, will also be incomplete lekta and hence incorporeal. And this is the position usually taken by commentators, with some uneasiness, though, because we are also told that according to the Stoics nouns signify qualities (D.L. VII, 58), but also that qualities, at least those of bodies, are themselves corporeal. So there seems to be an incoherence.

In looking for a solution to this problem it is useful to notice that in the many lists of kinds of lekta the only incomplete lekta ever to be mentioned are predicates (cf. e.g. D.L. VII, 43; 63; Plut. *De comm. not.* 1074D), that these are syntactical predicates, and that this is what we should expect if we consider the metaphysical function of lekta referred to above. The Stoics want to account for the fact that Socrates is wise (which is considered by them as a complete lekton), and they think that to account for this we need more than the two entities Socrates and wisdom. We also need the predicate of being wise. But if we have the predicate of being wise, or that x is wise, and if we have Socrates there is no need to postulate an additional item to correspond to Socrates. Socrates and the incomplete lekton that he is wise will suffice to make up the complete lekton that Socrates is wise. There is also no need to assume that since facts are not bodies no constituent part of a fact can be a body. Hence, so far there is no reason to think that cases are incorporeal. Quite the opposite, on the basis of what has been said, we should expect cases to be corporeal. And that, of course, would fit the testimony perfectly which says that nouns signify qualities, that is, according to the Stoics, corporeal entities.

But I think there is not only good reason to believe that the apparent contradiction can be solved by assuming that cases as opposed to predicates are thought to be corporeal, there is more direct evidence that the Stoics did regard them as corporeal.

To understand this evidence one has to take into account that Platonists (some claim even Plato himself) assumed two kinds of forms, the transcendent forms or ideas (ἄυλα εἴδη) and the immanent forms as they are found in concrete objects (ἔνυλα εἴδη). There is the transcendent form wisdom and, in addition, the form wisdom, a realization of the idea, which is to be found in a concrete individual, e.g. Socrates' wisdom. The second kind of form corresponds, it appears, to the Aristotelian forms of the Peripatetics. Both Platonists and Peripatetics agreed that these forms in themselves are incorporeal, though they are embodied, whereas the Stoics believed such qualities or forms to be as corporeal as the bodies they are the qualities of.

Now, the evidence I have referred to seems to be evidence to the effect that cases in the Stoic system correspond to the embodied forms of the Platonists and the Aristotelians. Stobaeus in a notoriously obscure and difficult passage (Ecl. I, p. 136, 21ff. W.) says that according to the Stoics "concepts (ἐννοήματα) are ... quasi-somethings (ὡσανεί τινα; cf. D.L. VII, 60; Origenes In Joh. II, 13, 93). They are called ideas by the ancients. For the ideas are ideas of the things which according to the concepts fall under them ... Of these (sc. ideas) the Stoic philosophers say that they do not subsist and that we participate in these concepts, whereas the cases, which they call appellatives, we do possess (τυγχάνειν)". A closely related passage is one in which Simplicius compares Stoic and Platonistic terminology (In cat. 209, 12–14): "the concepts they call participibilia (μεθεκτά) because they are participated in, and the cases they call possessibles (τευκτά) because they can be possessed (τυγχάνειν), and the predicates they call accidents". Finally a third passage should be mentioned: Clement, Strom. VIII, 9, p. 97, 6–7. Here it is said that the case is something which the concrete object possesses (τυγχάνει).

It appears from these three passages that there are items, namely the cases, of which both Stoics and non-Stoics say that they are somehow possessed (as opposed to participated in) by objects in this world. In this respect they are contrasted with the ideas which things in this world only participate in (but do not possess). It seems clear, then, that these cases are or correspond to the embodied forms of the Peripatetics and the Platonists, i.e. they are qualities of some sort.

There is some apparent counterevidence which it now should be easier to deal with. There is first of all the famous passage (A.M. VIII, 12) in which Sextus Empiricus explains that the Stoics distinguish between the utterance, the lekton and the external object the utterance is about and that they make the lekton the primary bearer of truth and falsehood. As an example Sextus gives 'Dion'. Hence, interpreters naturally have assumed that the Stoics distinguished between (1) the reference of a term, in our case the person Dio, (2) its meaning or sense, an incorporeal lekton, and (3) the word 'Dion', But unfortunately 'Dion' is not an example of something which is true or false, and so it is clear that something has gone wrong with Sextus' report (it is not a simple slip on Sextus' part; for cf. *A.M.* VIII, 75). To judge from parallel passages (e.g. Seneca Ep. 117, 13) the claim must have been rather that in the case of an utterance like "Dio walks" we have to distinguish between (1) the expression, (2) the corresponding lekton that Dion walks (it is this which is true or false), and (3) the external object we are talking about, Dio.

I am all the more inclined to think that we have to treat the passage in Sextus in this way as in one respect his report fits the interpretation we have given so far extremely well. Sextus says that the Stoics call the external object 'τὸ τυγχάνον' (cf. e.g. Philoponus In an. pr. 243, 2 and Plut. Adv. Colot. 1119F). Translators and commentators had some difficulty with this expression. The explanation for this use should be clear now. 'τυγχάνειν' was the expression used in the three passages quoted above to characterize the relation between cases or qualities and the objects which possess these qualities. The Stoics, then, call the external object 'τὸ τυγχάνον' because it is the object which possesses the quality or case signified by the noun.

If this is correct it is clear that the common view according to which the Stoics distinguished between the meaning and the reference of terms and posited lekta as the meanings is quite inadequate. The term 'horse' signifies the quality characteristic of horses. If in addition to this there is something faintly like the meaning of the term it is the concept. But of this the Stoics repeatedly (e.g. in the passage quoted above from Stobaeus) say that it does not exist, that it is not even a something but only an as-if-it-were something. Hence it cannot be a lekton since lekta are somethings. Moreover it seems that the Stoics did not make use of these concepts in their theory of meaning. And hence it would seem that they thought that all that was needed for such terms as 'horse' (appellatives and names) was the corresponding corporeal quality. Similarly for predicate-phrases like 'runs' we do not have two kinds of items corresponding to them but only one, the respective incomplete lekton. It is only in the case of utterances about something that we have to distinguish three kinds of items, the utterance, the lekton, and the external object.

I think it would be misleading to say here 'the object the statement or utterance is about' or 'the subject' instead of 'the external object'. At least some clarification is needed. For from what has been said it is clear that in a sentence with a proper name or a common noun in subject-position the subject-expression does not signify a subject of which the predicate has to be true if the sentence is to be true, to use Aristotelian language. There only has to be such an object which has the quality signified by the name or noun for the statement to be true. Hence there would seem to be room for a distinction between that which the statement is about, i.e. that which is signified by the subject-expression, i.e. the respective quality, and an object thus qualified, of which the predicate has to be true if the statement is to be true. There might be a parallel here to Aristotle's treatment of unquantified general propositions (the so-called 'indefinite' ones) in the *Categories* and *De interpretatione*. According to that account 'ἄν-θρωπος τρέχει' would have the species or universal 'man' as its subject (ὑποκείμενον), but it would be true in virtue of the fact that there is some particular of which the predicate is true. In any case later grammarians were puzzled by the Stoic assumption that names and common nouns signify qualities rather than something qualified in a certain way. And their attempts to modify the Stoic definitions resulted in a considerable variety of partly rather obscure definitions which gave rise to further problems. The introduction of the term 'substance' (οὐσία) into the definition, e.g. would raise the question whether substantial objects or essences were referred to, and in either case, but particularly in the latter, a problem would arise how adjectives would fit the definition (they so far had been treated as appellative nouns).

So much about what Sextus Empiricus has to say in *A.M.* VIII, 12 and some of the problems raised by this passage. But there is more evidence against our suggestion that cases, for the Stoics, are not incorporeal lekta, but corporeal qualities. The counter-evidence consists of two remarks in Clement's Stromateis, separated by only a few lines. Clement says that a case is incorporeal (VIII, 9, p. 97, 6–7) and that it is agreed to be incorporeal. This testimony surely should settle the matter against us.

Fortunately the matter is more complicated. It is not difficult to explain how Clement's remark, that a case is incorporeal, could be compatible with our view that cases for the Stoics are corporeal. Starting in the first century B.C. attempts were made to synthesize the doctrines of the rivalling schools, and all later Greek philosophy is heavily influenced by this syncretism. Hence we should not be surprised to find the term 'case' used by non-Stoics who would claim, though, that cases are incorporeal because they do not share the Stoic view that qualities are bodies. This was still a debated question in Clement's time as we can see from the fact that Galen (?), his contemporary, wrote a short treatise on the question (ed. Kühn vol. 19). An example of such a non-Stoic use of 'case' may perhaps be found in a fragment of Iamblichus (Simpl. *In cat.* 53, 17; cf. 53, 9ff.). Now the philosophy of Stromateis VIII is highly syncretistic (as has been pointed out again recently by S. Lilla in his book on Clement). In the theory of meaning Clement follows the Peripatetics (cf. VIII, 8, p. 94, 5ff.). Hence we should not be surprised to find that he uses the term 'case' but claims that cases are incorporeal because he thinks that qualities are incorporeal.

Much more difficult to deal with is his other remark, made a few lines earlier (p. 97, 3), that cases are agreed to be incorporeal. Part of the problem is that the text and its argument are so difficult to follow that editors have been tempted to change its reading in several places. But whichever interpretation we adopt it seems to be clear that the notion of a case with reference to which Clement here says that cases are agreed to be incorporeal cannot be the Stoic one in question. For the remark that cases are agreed to be incorporeal is made with reference to a remark three lines earlier (96, 26), that to be cut is a case, and a remark two lines earlier (97, 1), that for a ship to come into being is a case. These items certainly would be agreed to be incorporeal. But they would not be considered as cases by the Stoics. Hence Clement's notion of a case which not only covers the case of a house (cf. 97, 7) but also the two examples just mentioned must be different from the Stoic notion. For this reason and because the items with reference to which Clement says that cases are agreed to be incorporeal are in fact agreed to be incorporeal by the Stoics I do not think that Clement's remark can be used as evidence against the suggestion that Stoic cases are corporeal.

Given this, we may try to solve our first problem: why is the nominative called a case? It should now be clear that the problem arose because two quite different terms were conflated, since the same word '$\pi\tau\tilde{\omega}\sigma\iota\varsigma$' or 'case' was used for both and since both terms were used to refer to grammatical cases. There was the term of Aristotelian origin (cf. *De int.* 16a32ff.) according to which the word in the nominative is the noun whereas the word with the genitive inflection is not the noun but a case since it is thought of as falling from or being derived or inflected from the word in the nominative. Hence in this tradition the nominative could be called a case only by courtesy. But the Stoic case is called a case, as we are told in the passage I quoted from Stobaeus, because it falls under a concept or an idea. And since, of course, it falls under this concept quite independently of whether it comes in the nominative or the genitive the nominative for the Stoics is as much a case as any other (for possible traces of the correct explanation cf. Ammonius *In de int.* 43, 9–10).

Besides elementary predicates and cases the Stoics distinguished other kinds of elements of lekta. But I will not go into the matter here. Nor will I here discuss the reconstruction of the syntax of lekta. All I am concerned with now is to point out that the Stoics had something like a syntax of lekta and that a good part of the traditional grammatical terminology is derived from the Stoic specification and characterization of the elements of lekta whose combination is studied in the syntax of lekta.

This raises the question how these terms used or introduced to characterize items on the level of lekta came to be applied to items on the level of expressions. The answer, it seems, will have to take into account at least three factors. Apparently the Stoics believed that the relation between lekta and the sentences used to express them is basically simple and rational though disturbed over the ages by careless usage and tendencies towards abbreviation, euphony and the like. Chrysippus dealt extensively with such anomalies as they occur in the formation of words (cf. D.L. VII, 192). But the mere fact that such anomalies are singled out for special treatment appears to indicate confidence in the basic regularity with which features of the lekta and syntactical or morphological features of the expression correspond to each other. A good example of this confidence may be the following: the Stoics distinguished between individualizing qualities, i.e. features which make their owners the particular individuals they are, individual essences so to speak, and common qualities. But they were not content to distinguish correspondingly between proper names and general nouns (appellatives), they tried to confirm this distinction by showing that proper names are inflected differently from general nouns (*Schol. in D.Th.* 214, 19ff.; 356, 27ff.; Charisius p. 80, 1ff. Barwick). In a highly inflected language like Greek correspondences in gender, case, and number, e.g. would be fairly conspicuous, anyway. And hence it would be natural for the Stoics to use the same terms to refer to a feature of the lekton and the corresponding feature of the expression, as in fact we see they did (cf. 'ἐνικόν' and 'πληθυντικόν' in D.L. VII, 192 and Chrysipp's *Logical Investigations*). Secondly most philosophers and even some Stoics (S.E. *A.M.* VIII, 258) rejected the postulation of lekta. Hence they would be inclined to transfer the use of Stoic terms they found useful, where possible, to expressions. Thirdly, partly as a result of the two factors mentioned, considerable confusion seems to have set in. This may have been aided by the fact that the expression 'lekton' itself is ambiguous and was in fact used by the influential grammarian Chaeres for expressions as opposed to what they signify (S.E. *A.M.* I, 76; 77; 78), a use we then also find, e.g. in Apollonius Dyscolus (*De pron.* 59, 5–6; *De adv,* 158, 20; *De coni.* 233, 2ff.). Hence one finds that the two levels, so carefully distinguished by the Stoics in theory, constantly are confused even in good sources like Plutarch who in *Q.P.* (1009 C) claims that 'case' is the Stoic term for 'noun' and 'predicate' their term for 'verb'. This confusion is so widespread that some modern authors like A. C. Lloyd ('Grammar and Metaphysics in the Stoa', p. 58 in A. A. Long, *Problems in Stoicism*) have been misled into thinking that lekta are the expressions insofar as they are meaningful. A good Greek example of the terminological confusion is Ps.-Alexander on *Soph. El.,* p. 20, 27ff., where it is claimed that entities have to be divided into two classes, λεκτά and τυγχάνοντα, i.e. expressions and the things referred to by them. Hence there will be no shortage of explanations for the transfer of terms from lekta to expressions and the resulting confusions.

Let us now turn to the second part of Stoic dialectic which is concerned with voice and with expressions. To judge from the short survey given by Diocles Magnes in D.L. VII, 55–62, the Stoics under this heading dealt with a wide variety of topics, and it is not at all clear how these topics would form a systematic unity. But it is fairly clear that those of relevance for us are the first three, i.e. the sections on sounds, the parts of speech, and the virtues and vices of speech.

The idea behind this division seems to be fairly clear. In this part of dialectic we are concerned with the utterances we make, or expressions we use, when we claim, order, ask, express a wish or the like. These utterances may be regarded under two aspects: they may be regarded just insofar as certain sounds are articulated, or they may be regarded insofar as they are used to do or to express something. The first of the three parts mentioned clearly corresponds to a study of utterances under the first aspect. It is obvious why such a study would be of interest or relevance. To mention a less conspicuous reason one may point out that some Stoics insisted that one should also judge poetry by its mere sound (cf. Crates fr. 82 Mette). In the following I will not go into the details of Stoic phonology. The second part concerning the parts of speech corresponds to the study of utterances insofar as they are made to say something. This is less than obvious, and hence I will return to this point shortly. The section on the virtues and vices of speech, finally, is of traditional origin. Theophrastus, e.g. had dealt with the subject, and the Stoics show themselves to be influenced by him. This third part is of interest here insofar as it reveals to us a part of the motivation which made the Stoics go into technical grammar: the Stoic wise man is absolutely infallible, infallible also in any of the ways one may go wrong in speaking; to make him immune against such mistakes he has to acquire the relevant knowledge.

Given our initial rough characterization of traditional grammar one may wonder what happened to morphology and syntax. A consideration of this question will also help us to come to a better understanding of the second part concerning the parts of speech and the dominating role it plays in ancient grammar. The doctrine of the parts of speech plays such a dominating role that treatises on technical grammar very often are nothing more than treatises on the parts of speech. It is often maintained that syntax was only developed long after the Stoics and Dionysius Thrax (Steinthal, *Geschichte der Sprachwissenschaft*, vol. 2, p. 393, and Robins, *Dionysius Thrax.* ... , p. 102, e.g. claim that it is the work of Apollonius Dyscolus), and it is certainly significant that even in later times not many treatises on syntax were written: there are, besides Apollonius Dyscolus, the last two books of Priscian's Institutiones, and then the Byzantine tracts by Arcadius, Michael Syncellus, and Gregory of Corinth, though Priscian's remarks (Inst. XVII, 1) suggest that he had other authors to rely on besides Apollonius Dyscolus. And so we also have to look for an explanation of this phenomenon.

It seems to me that at least part of the explanation may be that the Stoics had some kind of syntax after all, but pursued it in a peculiar way. They certainly used the notions of concord, governance, and order (cf. e.g. D.L. VII 59; S.E. *A.M.* VIII, 90; Apuleius *De int.* p. 177, 27 Thomas). Apollonius Dyscolus in the introduction to his syntax (I, 2, p. 2, 10–3, 2; cf. also Priscian XVII. 3) tells us that to each word there corresponds an intelligible item (νοητόν) such that, in a way, it is these intelligible items which are the elements of the meaningful sentence. It is by combining the words that we combine the intelligible elements in such a way as to get a meaningful sentence. And we get a complete or perfect sentence (αὐτοτελὴς λόγος) if the intelligible items fit each other or, as Apollonius puts it, if the intelligibles have the required καταλληλότης. "καταλληλότης" is the standard Greek term, used by Apollonius and others, for syntactical propriety.

These remarks by Apollonius are easily translated into Stoic language, expecially since a good part of the terminology of the passage – including that for syntactical propriety (cf. D.L. VII, 59) – is of Stoic origin anyway. In translation Apollonius would say: corresponding to each word there is an element in the lekton; in putting the words together we put the elements of the lekton together, i.e. construct a lekton. Whether we get a syntactically proper sentence depends on whether the lekton we construct satisfies the syntax for lekta.

So it may turn out that the reason why we would look in vain for a

syntax in the part of dialectic concerned with expressions is that even
for Apollonius the syntax of sentences is still basically the syntax of the
corresponding lekta, except that Apollonius has replaced the Stoic
lekta and their elements by something like our meanings which he
refers to as intelligibles.

But the connection between Stoic syntax and Apollonius' approach
to syntax is even closer. So far we have talked as if the Stoics had just
been concerned with something like a syntax for lekta. But a closer
look at the evidence shows that the Stoics also were concerned with
the question how we use words such that by combining these words in
a certain way we get a lekton which satisfies the laws of the syntax of
lekta. Dionysius of Halicarnassus in his 'Composition of Words', e.g.
reports that Chrysippus wrote two treatises on the syntax of all sorts of
propositions (p. 22, 14ff. Usener-Radermacher). Among these propositions he lists ambiguous propositions. Now it is sentences, but not
propositions which according to the Stoics could be ambiguous. Hence
Chrysippus in these treatises on syntax must also have talked about the
composition of sentences to express the various kinds of propositions.
How this could be so we can see from many passages; e.g. from the
description of molecular propositions in Diogenes Laertius. There
(VII, 71–72) we are told that a disjunctive proposition is formed by
means of a disjunctive conjunction in such-and-such a way, and
similarly for other kinds of molecular propositions.

Passages like this also suggest an answer to the question why we
need a fairly elaborate doctrine of the parts of speech. Only such a
theory would provide us with the conceptual apparatus needed to
describe how we use words to put together sentences to express the
kinds of lekta distinguished by the syntax of lekta. That this is so is put
beyond reasonable doubt by the fact that Dionysius of Halicarnassus
gives as the title of the writings of Chrysippus he refers to "On the
syntax of the parts of speech". Matters become even clearer if we have
a look at the relevant part of the catalogue of Chrysippus' writings
(D.L. VII, 192–193). There the following four titles are listed in
sequence "On the elements of speech and what is said (5 books)" ,"On
the syntax of what is said (4 books)", "On the syntax and the elements
of what is said, to Philippus (3 books)" and "On the elements of
speech, to Nicias (1 book)". What these titles and their relative

position in the catalogue reveal, if we keep in mind that the 'what is said' in these titles refers to the lekta, is that the parts of speech or elements of speech were talked of and discussed in connection with the elements of lekta and their composition. Seneca (*ep.* 89, 9) mentions as one of the most important tasks of Stoic logic an investigation of the 'structura (sc. verborum)', i.e. the composition of words to form sentences. Similarly Epictetus (*Diss.* IV, 8, 12) attributes already to Zeno the view that one of the most important tasks of the philosopher is to determine the elements of speech, to characterize them, and to see how they fit each other (πῶς ἁρμόττεται πρὸς ἄλληλα). Hence it seems that the Stoic school right from the start had taken a strong interest in something like the syntax of sentences. And the question just is why among the sections of the second part of Stoic dialectic we do not find one on syntax.[3]

The very terminology for parts of speech seems to fit this picture very well. The parts of speech are the parts of a sentence which, in being combined, make up a sentence and in making up sentences make up speech. Chrysippus apparently wanted to stress their character of being elements of sentences and hence preferred the expression 'elements of speech' (Galen, *De Plat. et Hipp. dogm.* p. 673, 6 Müller; Theodosius Alex. p. 17 Göttling). Now, if we keep in mind that utterances are regarded under two aspects, as articulated sounds and as sentences, i.e. as utterances insofar as they express a lekton, then it becomes clear that the parts or elements of speech owe their very name to the fact that they are the elements which by being combined in a certain way will give us a lekton of a certain kind. All this, then, makes one inclined to think that the doctrine of the syntax of lekta together with the doctrine of the parts of speech is also supposed to serve the function of a syntax, and that one reason why the doctrine of the parts of speech plays such a prominent role is that it supplies us with the concepts we need to describe how one combines expressions to get a sentence and hence a lekton. And one reason why so few treatises on syntax were written may be that it was thought that this was basically a matter of the syntax of lekta and hence belonged to dialectic.

Again, I will not here consider the details of the doctrine of the parts of speech. As we should expect, the parts of speech are made to correspond to the elements of lekta. Even later Apollonius Dyscolus (*De pron.* 67, 6–7) will claim that the parts of speech are divided according to their meanings. Hence the so-called 'notional' character of the traditional definitions of the parts of speech. For the Stoics this is clear in the case of proper names, appellatives, and verbs, which are defined by them as corresponding to individualizing qualities, common qualities, and simple predicates, respectively. It is not as clear for the definition of the article which is defined as determining gender and number. And it even seems to be shown wrong by the definition of the conjunction as an undeclinable part of speech binding together the parts of speech (cf. D.L. VII, 58). Yet this exception in an important sense is only apparent. Behind it there seems to be a discussion in the Stoic school concerning the so-called expletive conjunctions, i.e. particles like 'now' in "well, now, let us do so-and-so" which are used a lot in Greek, e.g. to ensure proper sentence-rhythm. The discussion as reported by later authors concerns the question whether conjunctions have a meaning, i.e. whether to the conjunction in the sentence there always corresponds an element in the lekton. According to Apollonius Dyscolus (*De coni.* 248, 1ff.) the Stoic grammarian Chaeremo denied this with reference to the expletive conjunctions but explained that they nevertheless could be called conjunctions because of their morphology which they share with the 'true' conjunctions. This should help to explain the apparently deviant definition of the conjunction.

This dispute concerning conjunctions is of interest in other respects, too. It shows, e.g. that the Stoics were aware of the fact that the structure of the sentence is not determined exclusively by the structure of the underlying lekton. It also shows a concern for the morphology characteristic of the various parts of speech. This interest is not surprising. After all, it is not very helpful to be told that one forms a disjunctive proposition by means of a disjunctive conjunction unless one knows what counts as a conjunction and what, in particular, counts as a disjunctive conjunction. Hence attempts were made to distinguish the parts of speech and their sub-kinds and forms by their morphology and surface syntax. And hence, perhaps, one resorted in those cases where such distinctions could no longer be made, to lists of words, e.g. in the case of the various kinds of conjunctions and prepositions (cf. Dionysius Thrax p. 71, 1ff.).

So much for a very rough sketch of the parts of Stoic dialectic insofar as they are relevant for traditional grammar. I should also have dealt with the question to what extent the Stoics were really interested in an analysis of the Greek language in general (or at least some form of it). This question arises because so many of the grammatical distinctions the Stoics make may seem to be made out of a logical interest. One's suspicions could grow if one sees that Dionysius of Halicarnassus (*De comp. verb.* p. 22) says that when he wanted to treat composition he turned to the Stoics because of their renown in linguistics (λεκτικὸς τόπος), but that he was quite disappointed by Chrysippus' writings on syntax because they rather seemed to belong to dialectic. Similarly one might become suspicious when Apollonius Dyscolus in the introduction to 'On conjunctions' says that the Stoics have written a lot on the matter but warns against those grammarians who introduce Stoic terms and notions which are alien to or not useful for grammar.

To deal with this question adequately one would have to have a clear notion of the relation between logic and grammar in general and in the Stoics in particular. From what has been said it is clearly not their ignorance of or lack of interest in grammar which makes them fail to recognize grammar as a discipline separate from dialectic. But all I can do now is to point out that disparaging remarks like the ones referred to can easily be explained and that many of the grammatical distinctions made by the Stoics are not systematically connected with what we regard as the logic of the Stoics and hence have to be explained as due to an interest in the language. The remark by Dionysius is easily explained by the fact that he, as he says, is interested in rhetorical composition. And somebody primarily interested in style and rhetoric would regard a treatise on syntax, however grammatical in orientation it was, as dialectical. The grammarians in the Alexandrian tradition would not be too happy with Stoic grammar because they want to understand the language of Homer, the poets, and in general the classical authors, whereas the Stoics, though they are very much interested in Homer, primarily want to construct or reconstruct the language the wise man would use. It later will become a common-place that Chrysippus and the Stoics try to tell even the Athenians how to speak Greek properly (Galen *De diff. puls.* 10, SVF II, 24 von Arnim; cf. S.E. *A.M.* VIII, 125–126;

Cicero *De fato* 8, 15). But if the Stoics tried to 'legislate' the use of language, as their opponents put it, they must at least themselves have thought that they knew something about the way language works.

It is time to return to our argument concerning the origins of traditional grammar. We saw that a closer look at the evidence shows that it is very unlikely that traditional grammar started with Dionysius Thrax or his Alexandrian predecessors. But we also felt uncertain whether we should attribute the origins of traditional grammar to the Stoics. And one reason why we felt uncertain about this was that the Stoics may not have done more than to make some, perhaps important, but nevertheless incidental contributions to grammar.

But from what has been said it should be clear that Stoic contributions to grammar were not incidental. They were made as part of a larger theory which also served the purpose of accounting for utterances or sentences of Greek. Moreover we have seen that some grammatically relevant parts of this larger theory had been worked out by the Stoics in considerable detail, that a very large portion of the notions of traditional grammar were incorporated into or made their first appearance in this theory, and that the work of later ancient grammarians can be regarded as taking up well-defined parts of the Stoic project of a study of the Greek language as outlined above. We have also seen that we may be able to explain the curious prominence of the doctrine of the parts of speech in later grammar if we assume that this grammar developed out of Stoic dialectic. Hence the claim that traditional grammar is of Stoic origin seems to be quite promising. With this in mind it should be easier to evaluate some further evidence.

There is general agreement that among the Stoics with an interest in grammar Diogenes of Babylon was of particular influence. In the short survey in Diogenes Laertius he is referred to eight times within the five paragraphs we are concerned with (VII, 55–59). Now among the pupils of Diogenes of Babylon there are two of importance for us. One of them is Crates who went to Pergamum to establish a 'school' there which rivalled that of the Alexandrian philologists. Crates spent some time on a mission in Rome where he lectured on grammar. If we follow Suetonius in his *Grammarians and Rhetoricians* (Chap. 2) it were these lectures by Crates, "the equal of Aristarchus", which started grammar in Rome. Of course, 'grammar' here has to be understood in the wide sense. But we can be fairly certain that this for Crates included the type of Stoic grammar we have discussed above. For Sextus Empiricus (*A.M.* I, 79) tells us that Crates required of a grammarian (whom Crates called a 'critic') that he should be familiar with logic. It seems plausible to assume that Crates had in mind that the grammarians of the Alexandrian kind lacked the knowledge of Stoic logic including the kind of grammar discussed above which is required to do philology properly. And this may get some confirmation from the fact that a later Cratetean, Tauriscus, called the technical part of grammar 'the logical part' (S.E. *A.M.* I, 248). In any case, Roman technical grammar was centered around the doctrine of the parts of speech, sometimes preceded by a phonology and sometimes followed by a part on the virtues and vices of speech, i.e. it seems to follow the basic structure of the relevant part of Stoic dialectic. Perhaps most telling is the way the Roman art of grammar preserved in some instances a part of the Stoic techne we have not referred to at all because it does not belong to traditional grammar (cf. K. Barwick's Remmius Palaemon). We know that the Stoics in the theory of expressions or utterances at least sometimes also dealt with the notions of the definition, the species, and the genus. Like a fossil we still find these topics dealt with in a few lines by Charisius (p. 192, 20–193, 2 Barwick), with no connection to the preceding or the following. Later Roman grammarians know even less what to do with this fossil, and hence Diomedes, Maximus Victorinus, and Audax at least keep a definition of 'definition' in an inobstrusive position. Diomedes even manages to put in definitions of 'genus' and 'species' in some other place (326, 30–35). In many ways, then, the Roman art of grammar suggests that it is of Stoic origin. Priscian, at the end of a long tradition in Latin grammar, tells us in fact that Roman writers on technical grammar tended to follow the Stoics (XI, 1, 1, p. 548, 12ff.).

Another student of Diogenes of Babylon was Apollororus, who afterward went to Alexandria to work with Aristarchus and who was to become, besides Dionysius Thrax, the most important pupil of Aristarch. We may assume that he was quite familiar with Diogenes' grammatical ideas, and the only thing we know about his views on technical grammar makes him side with the Stoics rather than with

Dionysius' *Techne* and later Greek grammar: he classified pronouns as demonstrative articles (Ap. Dysc. *De pron.* 5, 18–19). Now it is certainly not the case that the Alexandrians had to rely on oral reports to learn about the theories of Diogenes of Babylon. But it is equally difficult to believe that a distinguished scholar like Apollodorus, trained in Stoic logic, should not be of considerable influence on the Alexandrians in this respect. And hence it is tempting to see some significance in the fact that in all three cases in which we know that Dionysius at some time defended a point of view different from the one he took in the Techne he was siding with the Stoics and that at least on one of these points he was in company with Apollodorus. Like Apollodorus at some stage he thought of pronouns as demonstrative articles (Ap. Dysc. *De pron.* 5, 18–19). In addition, like the Stoics, he defined the verb as signifying a predicate (*Schol. in D.Th.* p. 161, 6; Ap. Dysc. *Fragm.* p. 71, 29ff.), and finally, like the Stoics, he at some point counted proper names and common nouns as two parts of speech (*Schol. in D.Th.* 160, 26–28). If we want to maintain the authenticity of Dionysius' Techne the easiest explanation of the facts may be that Dionysius, perhaps through Apollodorus, got interested in Stoic grammar, but then modified it; modified it under the influence of the enormous experience and feeling for the Greek language the Alexandrians had, at least up to the time of Aristarchus. Parts of Dionysius' Techne in any case cannot be understood except as modifications of Stoic doctrine. The classification of conjunctions, e.g. in its content and its terminology reflects the Stoic doctrine at a certain, not fully developed stage. The lines on syllogistical conjunctions clearly presuppose Stoic rather than Aristotelian logic, and a Homeric scholar would hardly introduce a class of syllogistical conjunctions in the first place unless he was following a Stoic source fairly closely.

One could proceed to list more details of this sort. One certainly should mention that in 'De congressu eruditionis causa' Philo tries to make us believe that grammar had its origin in philosophy, Stoic philosophy as his terminology shows (sections 146–150). But I think we are now in a position to see that and in what sense Philo may be right. The Stoics certainly did not write grammars like Kühner-Gerth's or Schwyzer-Debrunner's. But on the other hand they did something their philological and philosophical predecessors failed to do, however much they may have contributed to grammar. It was the Stoics who first tried to develop a theory on the basis of which one would know what could pass as correct Greek. To the extent that later ancient grammarians took up the subject where it had been left by the Stoics and established the tradition which was to last to these days we can talk of the Stoic origin of traditional grammar. Why and how grammar was separated from logic or dialectic and came to fall within the province of the grammarian should deserve an account of its own.

NOTES

* I would like to thank Professor Julius Moravcsik for his kind discussion of points raised in this paper and related topics.
[1] The standard Greek terms are 'εὐκτική' and 'προστακτική' (cf. D.Th. 47, 3). Of these the second comes straightforwardly from the Stoic 'προστακτικόν' (cf, D.L. VII, 67). The first comes from 'εὐκτικός (sc. λόγος)', a term the Peripatetics preferred to the Stoic 'ἀρατικόν' (cf. Ammon. In de int. 2, 27), presumably under the influence of Aristotle, De int. 17a4, who in turn may have followed Protagoras' 'εὐχωλή' (D.L. IX, 53), but avoided the epic form.
[2] The standard Greek term is 'κλητική' (cf. D.Th. 31, 6). But D.Th. also gives as an alternative 'προσαγορευτική' (32, 1; cf. Choerob. 11, 3; Prisc. V, 73, p. 186, 1). 'κλητικός' is the Peripatetic equivalent of the Stoic 'προσαγορευτικόν' (cf. Ammon. In de int. 2,7) which is their standard term to refer to an address (cf. D.L. VII, 67).
[3] Syntacticality is, of course, referred to in the part on the virtues and vices of speech in the definition of a solecism (cf. D.L. VII, 59). But to judge from grammar treatises with a section on the vices and virtues of speech and the extant treatises on solecism it seems that it was not under this heading that syntax was developed systematically.

WILLIAM R. SHEA

GALILEO AND THE JUSTIFICATION OF EXPERIMENTS

Everyone is enough of an empiricist to believe that we learn from experience and no one is so far removed from rationalism as to deny that ideas play a vital role in the theories we construct about the world. But it is only too easy, and perhaps too tempting, for philosophers with a pronounced empiricist or rationalist bias to caricature the position of their opponents and make them appear as holding ludicrously simplistic views about the nature of scientific knowledge. In fact, any philosophical position worth its salt has a built-in flexibility or a power to accommodate, sometimes with surprising comfort, theses that seem central to rival theories. The Leaning Tower experiment, however fictitious, is equally well explained by rationalists or empiricists. Facts, however hard or obdurate, have a way of lending themselves to varying interpretations. Properly marshalled, they can be enlisted to serve any good philosophical cause.

Of course, one does come across extreme statements but these are usually found in popular books or among the *obiter dicta* of professors who, despairing of getting their undergraduates to grasp the complexity of the problem, vow that they shall, at least, remember one illuminating falsehood. "It is better", they say, "to have students believe that Galileo never performed experiments rather than have them profess with glib positivistic complacency that he dropped balls from the Tower of Pisa and that their thud made the opponents of scientific progress speechless to the present day." Such drastic remedy may be intended to counteract the influence of Oliver Lodge's *Pioneers of Science*, first published in 1893, and still going strong in a paperback edition. Lodge not only tells us what Galileo's method was like, he also draws a moral, and issues a warning:

Galileo was not content to be pooh-poohed and snubbed. He knew he was right, and he was determined to make everyone see the facts as he saw them. So one morning before the assembled University, he ascended the famous leaning tower, taking with him a 100 lb shot and a 1 lb shot. He balanced them on the edge of the tower, and let them drop together. Together they fell and together they struck the ground.

Butts and Hintikka (eds.), *Historical and Philosophical Dimensions of Logic, Methodology and Philosophy of Science*, 81–92.
Copyright © 1977 by D. Reidel Publishing Company, Dordrecht-Holland. All Rights Reserved.

The simultaneous clang of these weights sounded the death-knell of the old system, and heralded the birth of the new.

But was the change sudden? Were his opponents convinced? Not a jot. Though they had seen with their eyes, and heard with their ears, the full light of heaven shining upon them, they went back muttering and discontented to their musty old volumes ⋯ We need scarcely blame these men; at least we need not blame them overmuch. To say that they acted as they did is to say that they were human, were narrow-minded, and were apostles of a lost cause ⋯ Conduct which was excusable then would be unpardonable now, in the light of all this experience to guide us. Are there any now who practically repent their error and resist new truth? Who cling to any old anchorage of dogma, and refuse to rise with the tide of advancing knowledge? There may be some even now.[1]

We cannot read these words without a smile and perhaps a touch of nostalgia for an age when men knew not only how science had originated but where it was taking us. We cannot divorce our intellectual interests from the assumptions of our age anymore than we can step out of its technological limitations. As long as men believed that the 17th century marked a break with the past, that it heralded the dawn of a new intellectual period, it was normal to seek to clarify the nature of the scientific method by contrasting it with the method that prevailed before. One of the most plausible explanatory models was devised by Ernst Mach who saw the Galilean breakthrough in the discovery of the law of inertia. In this, he had an illustrious forerunner in the person of Isaac Newton who took it for granted that Galileo had preceded him in formulating the First Law of Motion. Mach could not positively assert that Galileo discovered the law by making experiments but he was in no doubt "that he tested the law experimentally" for "Salviati, chief advocate of Galileo's doctrines in the *Discorsi*, assures us of his repeatedly taking part in experiments, and describes these experiments very accurately."[2] Mach's Teutonic rigour made no allowances for Latin exuberance.

When the historical studies of Emil Wohlwill made it clear that Galileo's concept of inertia had the troubling feature of circularity, Mach was not unduly perturbed. What Galileo did not say but could have said, had he been fully aware of the logical consequences of his own position, became what really counts. However deficient Galileo's explicit utterances may have been he was, at least implicitly, in possession of the correct doctrine.

The best way of dispensing with a solution that is considered unsatisfactory is to attack the question that it was meant to answer.

This is what Pierre Duhem did when he argued, in a series of learned monographs, that the scientific revolution was the outcome of the continuous growth within the Aristotelian tradition culminating in the notebooks of Leonardo da Vinci. Duhem had to get around the problem that Leonardo's notes were not published before 1797 but he did this quite elegantly by drawing attention to the availability of the manuscripts to many 17th century scientists including Galileo.

Duhem's claims for the influence of Leonardo are now seen to have been grossly exaggerated but his more general thesis that the scientific revolution really occurred in the Middle Ages is still alive although I would be the last one to give it a clean bill of health. Recent studies by A. C. Crombie and William Wallace have shown how much Galileo owes to the Jesuits of the Roman College but it would be rash to see in their broadly based Aristotelianism the fountainhead of his philosophy of nature.

When Mach's phenomenalism came to America it adapted itself to the new environment and became operationalism. In time Galileo was duly summoned before the California Grand Jury. Edward W. Strong asked: "Although there is evidence in Galileo of a belief in mathematical realism, is Galileo an exception to the operational approach to science?"[3] Needless to say, Galileo was found innocent of such a breach of scientific methodology. In our own day, Stillman Drake has argued, in a much more subtle and sophisticated way, that Galileo discovered, albeit implicitly, the principle of inertia and that he was, albeit unwittingly, something of an operationalist.[4]

Why, we may ask, this fascination with Galileo? Could it be that having failed to produce a full-fledged account of his method he issued an open invitation, as it were, to philosophers of science looking for a suitable testing ground for their theories? "On my philosophy", I seem to hear, "we can, at last, understand why Galileo did (or did not) do this, that or the other." The nature of the scientific revolution need not be that mysterious after all!

Mach and Duhem represent two ways of looking at Galileo. A third approach is exemplified in the writings of Alexandre Koyré. In his *Etudes galiléennes*, he showed that Mach's interpretation would simply not do and that Duhem's arguments, however ingenious, merely disguised the problem it was meant to dissolve. Galileo may have been the heir of mediaeval mathematicians but he certainly managed their estate in a way that never entered their heads. The novelty in Galileo's position, Koyré argued, lies mainly in his commitment to a different outlook on nature. He may have used techniques that were devised by mediaeval predecessors, but he did so with radically different presuppositions. Whereas they operated on the general Aristotelian assumption that mathematics was an abstract science, whose explanatory power fell short of accounting for the empirical features of this world, Galileo maintained that only mathematics would disclose the true features of physical reality.

In English-speaking countries, views akin to Koyré were broadcast in E. A. Burtt's influential *The Metaphysical Foundations of Modern Science*, and after the war, in the lively and lucid historical outlines of Sir Herbert Butterfield, Rupert and Marie Hall, Stephen Toulmin and a host of recent writers. Sir Karl Popper, with characteristic vigour, made it repeatedly plain to those who would hear him, that brute facts are perfectly dumb and that all meaningful experiments are answers to prior questions.

One cannot mention Sir Karl Popper and Ernst Mach without being reminded of the accommodating character of any reasonably sophisticated theory. I am struck, for instance, by the similarity between these two utterances. The first is Sir Karl's, the second Mach's:

... observation statements and statements of experimental results, are always interpretations of the facts observed ... they are *interpretations* in the light of theories.[5]
Without some preconceived opinion the experiment is impossible, because its form is determined by the opinion. For how and on what could we experiment if we did not previously have some suspicion of what we were about?[6]

The quotation from Sir Karl Popper brings us to the year 1959 and it is interesting that a debate protracted over such a long period of time should not have moved any of the participants to attempt to repeat the experiments Galileo adduced as evidence for his theories. Koyré ruled most of Galileo's experiments out of court on the following grounds. First, Galileo himself warns that his experiments would only work under ideal conditions, such as having perfectly round balls and frictionless surfaces.[7] Secondly he often speaks of experiments as illustrating rather than confirming a given theory.[8] I should add, however, that Koyré laid great store by the experiments of Galileo's contemporaries and that he was particularly impressed by Mersenne's failure to attain the same results as Galileo when he performed the inclined plane experiment described in the *Two New Sciences*.[9]

The first modern attempt to reproduce the inclined plane experiment under the conditions outlined by Galileo was made by Thomas B. Settle of the Brooklyn Polytechnic.[10] He found that Galileo could have obtained empirical results as satisfactory as those he claimed. This outcome was hailed by empiricists as a vindication of their cherished belief that Galileo had, all along, been one of their own. In Stillman Drake's words: "Galileo's well-known statements about his experiments on inclined planes were completely vindicated."[11]

Now both Koyré's dismissal of Galileo's experimental claims and Drake's enthusiastic support for their reliability seem to me to be excellent illustrations of 'thought experiments'. Galileo speaks of perfectly round balls rolling down perfectly smooth surfaces. Koyré concluded that such experiments were beyond the art of experimentation, and, since Settle showed that Galileo could have performed the inclined place experiment, Drake concluded that he actually achieved Settle's results. But surely the most that we can claim for Settle's experiments, if they faithfully reproduce Galileo's, is that Galileo could have secured identical results but not that he necessarily did. *A posse ad esse non valet illatio*, not even for Galileo, although, as we shall presently see, he appears to have subscribed to a version of this principle.

In any event, Settle's experiments have now been called in doubt and the empiricists are again in danger of losing one of their players. In 1973, Ronald Naylor of the Thames Polytechnic noticed discrepancies between Settle's experiment and Galileo's own description, and he undertook to repeat the experiment with greater fidelity to Galileo's own words and a better awareness of experimental conditions in the 17th century.[12] Naylor found that Galileo could not have obtained anything as good as the results he claimed. If the experiment vindicates anyone it is Mersenne who marvelled at the discrepancy between his experimental results and those that Galileo published in the *Two New Sciences*.

Perhaps a word of explanation is warranted here. Settle achieved his results by rolling a ball down not the groove of the inclined plane but the edges of the groove. This greatly reduces the effect of spinning which robs the ball of much of its acceleration and, hence, leads to closer agreement with the law that relates distance traversed to time elapsed. But Galileo did not perform his experiment in this way. First, he refers to a relatively wide groove, namely "a little more than one finger in breadth" or between $\frac{3}{4}$ and 1 inch across (18 and 23 mm),[13] and, secondly, he states that he lined the grove with parchment which would not be necessary if the ball were not to be rolled down the groove.[14]

The mention of parchment has suggested to many readers of the *Two New Sciences* that Galileo first rolled a ball down an unlined groove, realized that there was too much friction and sought to reduce it. He knew that parchment was smooth and he used it to line the groove, thereby achieving the results that Mersenne marvelled at. Alas, this will not do. Using parchment actually reduces the speed of the ball. The reason is that parchment, being made of calf or sheep-skin, cannot be more than three feet long, and however carefully the ends are joined they cannot be made smooth enough to ensure unimpeded passage. If Galileo had made the experiment he would have realized that using parchment provides no advantage whatever. It is difficult to avoid the conclusion that Galileo performed a thought-experiment in which he imagined that friction would be eliminated if the groove were lined with parchment.

The *Two New Sciences* provide us with another striking instance of Galileo's 'experimental' method. I refer to his use of the pendulum to illustrate that bodies of differing weights fall at the same speed. Here is what Galileo says:

The experiment made to ascertain whether two bodies, differing greatly in weight will fall from a given height with the same speed offers some difficulty ... It occurred to me therefore to repeat many times the fall through a small height ... Accordingly, I took two balls, one of lead and one of cork, the former a hundred times heavier than the latter, and suspended them by means of two equal fine threads, each four or five cubits long. Pulling each ball aside from the perpendicular, I let them go at the same instant, and they, falling along the circumference of circles having these equal strings for semi-diameters, passed beyond the perpendicular and returned along the same path. This free vibration repeated a hundred times showed clearly that the heavy body maintains so nearly the period of the light body that neither in a hundred swings, nor even in a thousand will the former anticipate the latter by as much as a single moment, so perfectly do they keep step.[15]

The whole basis of Galileo's experiment is the supposed isochronism of the pendulum. Huygens saw that this could not be the case and Ronald Naylor is to be credited for showing that Galileo must have known that this was not the case. Naylor conducted a range of tests using brass and cork pendulum bobs and found that contrary to Galileo's description it is possible to detect a significant difference between the periods of pendula of identical length but of different bob material. Using two 76 inch pendula, one having a brass bob, the other one of cork, both swinging through an arc of 30°, the brass bob was seen to lead the cork by one quarter of an oscillation after only twenty-five completed swings. The time interval between the bobs passing the mean point after these twenty-five oscillations could easily be observed as it was about half a second.[16]

Again we might conjecture that Galileo used pendula with identical bobs and rashly generalized to pendula with different bobs. Galileo does describe in the Fourth Day of the *Two New Sciences* an experiment whereby he suspended two equal leaden balls from the same height and let them swing through different areas:

One through 80 or more degrees, the other through not more than four or five degrees ... if two persons start to count the vibrations, the one the large, the other the small, they will discover that after counting tens and even hundreds they will not differ by a single vibration, not even a fraction of one.[17]

In spite of Galileo's insistence that "after counting tens and even hundreds they will not differ by a single vibration, not even a fraction of one", the simple pendulum, even for modest arcs is not truly isochronous. This is well known. The interesting question is to what extent do Galileo's conclusions clash with observation. For pendula of two or three *braccia* long (between 40 to 60 inches), the length mentioned by Galileo to one of his scientific friends, the discrepancy is striking.[18] For longer pendula of four to five braccia (80–110 inches) mentioned in the *Two New Sciences*, there is a greater appearance of isochronism but this is largely an illusion produced by the greater magnitude of the size of the oscillation. Galileo does not, however, indicate any difference between long and short pendula.

If we now turn from an examination of the quality of the inclined plane experiment to the role that the experiment is meant to fulfill in the *Two New Sciences* we may perhaps gain some insight into Galileo's strategy. At the beginning of the Third Day which, as I have already indicated, is devoted to an analysis of accelerated motion, Salviati says:

We have decided to consider the phenomenon of bodies falling with an acceleration such as actually occurs in nature and to make this definition of accelerated motion exhibit the essential features of observed accelerated motion.[19]

The definition that is proposed is: Uniformly accelerated motion is that which, beginning from rest, acquires equal moments of swiftness in equal times. To this is added a postulate designated as *the only principle:* equal speeds are acquired in descent along different inclined planes if the planes are of the same height.[20]

Sagredo instructed by his *lumen naturale*, is willing to accept this Postulate, but Salviati describes an experiment that will increase the probability to something approaching necessary (that is, rigorous) demonstration. The experiment involves the familiar pendulum and its point is that the bob goes up to the same height from which it was released even if its path is shortened by placing a nail in its way. Sagredo is again willing to take the Postulate as demonstrated but Salviati restrains his enthusiasm:

For the present let us take it as a Postulate, the absolute truth of which will be established when we see other conclusions built on this hypothesis correspond to and agree perfectly with experience.[21]

This would appear to be a clear instance of Galileo's recognition of the hypothetico-deductive method. The truth of the definition and the postulate are to be established by the experimental corroboration of their inferred conclusions. A close search of the remainder of the Third Day, however, will reveal only one experiment that is adduced to show agreement between a deduced proposition and experimental evidence. This is the inclined plane experiment, and although Galileo claims to have many more which confirm his principles he adduces none other.

How are we to account for this singular behaviour? It is clear that Galileo was satisfied with rather less experimental support for his position than he implies. Ronald Naylor suggests that Galileo

did not feel it was always justified to expect precise agreement between observation and theory. Because of this he did not see why it was necessary to reject an essentially sound theory when complete agreement with observation was not found. If precise confirmation of a theory was not obtained, Galileo appears to have concluded that this was likely to be due to 'external' factors or defects in the experiment. If the pendulum did not appear completely isochronous, this was because of secondary effects. Galileo certainly never believed that the precise agreement between theory and measurement he indicated in the *Discorsi* (the *Two New Sciences*) was an essential requirement.[22]

Now this is a serious indictment for it presupposes that Galileo was not only disingenuous in proposing a method that could not produce the expected results but downright fraudulent in claiming to have secured evidence that was beyond his reach. A theory that is only partly confirmed by observation is 'essentially sound' only in the trivial sense of not being self-contradictory.

I believe we can understand what is happening if we do not assume that Galileo is following a procedure that foreshadows the hypothetico-deductive method. As Winifred Wisan has clearly seen, the principles of mechanics are not hypotheses for Galileo and the appeal to experience "is intended to render them evident, as is appropriate for a mathematical science, not to confirm them as proba-ble".[23] Galileo's new science of motion required assumptions that were neither self-evident nor immediately verifiable in experience.

Unable to claim that his principles were intuitively obvious and at a loss to account for their establishment through a more complicated procedure than was customary for a mathematical science, Galileo allowed his rhetorical gifts to make the facts bear witness to much more than was warranted. Ancient writers, whom he admired, had proceeded in this way. For instance, in Book III of his *Optics*, Ptolemy explains that knowledge begins with general principles that are "cer-tain and indubitable as seen by their effects and their internal consis-tency", for although "known in themselves" they are "made manifest by the appearances". One such principle is that the angle of reflection is equal to the angle of incidence and this is demonstrated by a single experiment. In other words, a basic assumption, illustrated by a simple experiment, is regarded as known. This characterizes Galileo's proce-dure in mechanics where he established his principles by *making them manifest by the apperances*, after which he assumes that necessary conclusions are derived by "rigorous mathematical deduction."[24] The difference, of course, is that Galileo's principles needed to be estab-lished by a process of confirmation that was more complex than anything that was customary and whose nature eluded Galileo himself.

Galileo was convinced that he had "necessary demonstration" not only
of the true nature of motion but of "the true constitution of the
universe". Alistair Crombie has recently indicated how much Galileo
owes to Aristotle and it seems clear that while he demolished Aristotelian physics, he remained an Aristotelian in his conception of the
nature of scientific knowledge.[25] He saw the "necessary demonstrations" of the *Posterior Analytics* as the hallmark of genuine science
which he found exemplified in Euclid and Archimedes. Neither he nor
his opponents realized that he had not only introduced new scientific
demonstrations but a new way of arriving at scientific interferences. In
the meanwhile, Galileo used what may be described as his version of
the *A posse ad esse* principle.

I remarked at the beginning of this paper on the accommodating
character of theories of scientific knowledge. Allow me to conclude
with a further instance.

The inclined plane experiment is usually cited as conclusive evidence
for the law that relates the distance fallen to the square of the time. In
the version given by Galileo this usually takes the form of the odd-
number rule or that the cumulative distances from rest are as 1, 3, 5,
7, ⋯ Surely such a progression could not be reconciled with the
classical impetus theory that Galileo had maintained in his youth and
that die-hard Aristotelians professed in universities. But as a matter of
fact, without denying Galileo's experimental results, Giovanni Battista
Baliani contended that a deeper analysis revealed that behind Galileo's
odd-number rule there lay the simpler and correct rule of the natural
numbers, 1, 2, 3, ⋯ for the spaces traversed in successive times. The
proof runs as follows: Take a distance actually traversed and divide it
into ten sections progressing in length as the natural numbers. Let the
first ten sections be traversed in a given time, the next ten sections in a
second equal time, and so on. The distance covered in the first time is
55 of the sum of the first ten spaces; the distance covered the second
time is 155, the sum of the 11th to the 20th space; and that in third,
255. But 55:155:255 are near to 1:3:5 and if we subdivide the
distance traversed into 100 parts instead of 10 we get an even closer
approximation namely: 5050:15050:25050. Hence, Baliani concludes, the spaces actually passed in very small intervals of time
can be as the natural numbers 1, 2, 3 although for the purpose of

measurement (we might say at the macroscopic level) the distance will appear as 1, 3, 5...

Baliani's "simple" conception of the underlying nature of motion, which was shared by other 17th century writers, did not survive in textbooks of physics, but whether this is because it was disproved or merely because it went out of fashion is not clear. In any event, it was not shattered by any of Galileo's experiments.

McGill University

NOTES

[1] Oliver Lodge, *The Pioneers of Science* [1893]. New York: Dover 1960, pp. 90–92. In his *De Motu*, written while he was in Pisa (1589–1592), Galileo mentions objects dropped from high towers, but he reaches the erroneous conclusion that the velocity is determined by the specific gravity of the body and that, hence, a ball of iron will fall faster than a ball of wool of the same weight.

[2] Ernst Mach, *The Science of Mechanics*, translated by Thomas J. McCormack. La Salle, Illinois: Open Court, 1960, p. 157.

[3] Edward W. Strong, *Procedure and Metaphysics* [1936]. Hildesheim: Georg Olms, 1966, p. 135.

[4] Stillman Drake, 'Galileo's Experimental Confirmation of Horizontal Inertia: Unpublished Manuscripts (Galileo Gleanings XX.11)', *Isis* **64** (1973), 291–305.

[5] Karl R. Popper, *The Logic of Scientific Discovery*. London: Hutchinson, 1959, p. 107, n. *2.

[6] Ernst Mach, *op. cit.* (footnote 2), p. 161.

[7] Galileo, Galilei, trans. by Henry Crew and Alphonso de Savio: [1914]. New York: Dover, 1953, p. 170. (In the *Edizione Nazionale* of Galileo's *Opere*. Florence: Barberà, 1890–1909, Vol. VIII, p. 206).

[8] This point is well made by Winifred L. Wisan, 'The New Science of Motion: A Study of Galileo's *De motu locali*', *Archive for History of Exact Sciences* **13** (1974), 122–123.

[9] Alexandre Koyré, 'An Experiment in Measurement', *Proceedings of the American Philosophical Society* **97** (1953), 222–237.

[10] Thomas B. Settle, 'An Experiment in the History of Science', *Science* **133** (1961), 19–23.

[11] Stillman Drake, *art. cit.* (footnote 4), p. 291.

[12] Ronald Naylor, 'Galileo and the Problem of Free Fall', *Brit. Journ. Hist. Sc.* **7** (1974), 105–134.

[13] Galileo, *Two New Sciences*, p. 178 (*Edizione Nazionale*, Vol. VIII, p. 212).

[14] *Ibid.*

[15] *Ibid.*, pp. 84–85 (*Edizione Nazionale*, Vol. VIII, pp. 128–129).

[16] Ronald Naylor, 'Galileo's Simple Revolution', *Physics* **16** (1974), 33.

[17] Galileo, *Two New Sciences*, pp. 254–255 (*Edizione Nazionale*, Vol. VIII, p. 277).

[18] See Galileo's letter of 29 November 1602 to Giuldobaldo del Monte, *Edizione Nazionale* of Galileo's *Opere*, Vol. X, p. 97.

[19] Galileo, *Two New Sciences*, p. 160 (*Edizione Nazionale*, Vol. VIII, p. 97).

[20] *Ibid.*, p. 169 (*Edizione Nazionale*, Vol. VIII, 205).

[21] *Ibid.*, p. 172 (*Edizione Nazionale*, Vol. VIII, p. 208).

[22] Naylor, *art. cit.* (footnote 16), p. 37

[23] Winifred L. Wisan, 'The New Science of Motion: A Study of Galileo's *De Motu locali*', *Archive for History of Exact Sciences* **13** (1974), 103–306.

[24] *Ibid.*, p. 125, n. 15.

[25] A. C. Crombie, 'Sources of Galileo's Early Natural Philosophy' in M. L. Righini Bonelli and W. R. Shea (eds.), *Reason, Experiment and Mysticism in the Scientific Revolution*. New York: Science History Publications, 1973, pp. 157–175.

[26] This is discussed in two recent articles by Stillman Drake, 'Free Fall from Albert of Saxony to Honoré Fabri', *Stud. Hist. Phil. Science*, **8** (1975), 347–366; 'Impetus Theory Reappraised', *Journal of the History of Ideas* **XXXVI** (1975), 27–46. Considerations analogous to those of Baliani were made as early as 1618 by Isaac Beeckman (Descartes, *Oeuvres*, edited by C. Adam and P. Tannery, 12 vols. and index, Paris: Vrin, 1956–1957, Vol. X, p. 61).

JOHN EARMAN

LEIBNIZIAN SPACE-TIMES AND
LEIBNIZIAN ALGEBRAS*

0. INTRODUCTION

In what is perhaps his only reference to Berkeley, Leibniz says:

The Irishman who denies the reality of bodies seems neither to offer suitable reasons nor to explain his position sufficiently. I suspect that he belongs to the class of men who want to be known for their paradoxes. (Loemker, 1970, p. 609.)

The first sentence embodies a misinterpretation of Berkeley, and the misinterpretation is particularly ironic in view of the fact that Leibniz, like Berkeley, offered a phenomenalistic reduction of the physical world.[1] The second sentence, however, has always struck me as being essentially correct. The 17th and 18th centuries were populated with intellectual giants who were struggling with the most fundamental problems in natural philosophy. In the foundations of mathematics and physics, Berkeley's contribution was to poke little paradoxes at the theories of the giants without offering positive alternatives and often without advancing the discussion at all.

The reason for raising this matter is that at one time it seemed to me that Leibniz played a Berkeleyesque role with respect to questions about the nature of space, time, and motion. Newton offered a whole explanatory framework while Leibniz offered objections. Leibniz also offered a few positive suggestions, but they are notoriously incomplete and slippery; even when he is at his clearest, it is never entirely clear what his position is or how it differs from the positions of his opponents. Nor, to my knowledge, has anyone showed how Leibniz's alleged insights can be incorporated into an explanatorily adequate scientific theory.

While I still retain part of this negative assessment of Leibniz's role, I now believe that his views on space and time lead to some profound issues which are important not only for an appreciation of historical

Butts and Hintikka (eds.), Historical and Philosophical Dimensions of Logic, Methodology and Philosophy of Science, 93–112.

controversies but which also have contemporary applications. Some of these issues have never been adequately identified much less sharpened in the course of the still lively debate between the Newtonians and the Leibnizians.

In order to separate Leibniz's genuine insights from his confusions, I will examine his views on space and time through the medium of space-time structures. I will also pay particular attention to the problem of motion because this was for Leibniz, as it was for all 17th century natural philosophers, the overriding problem in physics and because the crucial entity for the problem of motion is space-time.

1. MINIMAL LEIBNIZIAN SPACE-TIME

For Leibniz, spatio-temporal relations are well-founded phenomena. Since the doctrine of well-foundedness in particular and the mondology in general are beyond the scope of this paper, I will confine myself to the task of describing the structure and the phenomenal foundations these relations have according to Leibniz.[2]

In part, the task is an easy one. Leibniz is clear that there is a well-defined notion of simultaneity or coexistence; that each simultaneity sheet is E^3; and that the notion of duration or distance between simultaneity sheets is well defined. The difficult part comes in deciding what additional structure, if any, there must be to Leibnizian space-time. One plausible interpretation of a good deal of Leibniz's writings is to take him as saying that no other structure is present (see Stein, 1975). The following definition gives a somewhat more precise rendering of this interpretation.

DEFINITION 1. *Minimal Leibnizian space-time* L_1 is a triple (M, t, g). M is a differentiable manifold diffeomorphic to $\mathbb{R}^4$. t, the time matric, is a symmetric type $(0, 2)$ tensor field having the form $dt \otimes dt$ where t ('absolute time') is a smooth function. g, the space metric, is a symmetric type $(2, 0)$ tensor field having index 0 and nullity 1 with $g(dt, dt) = 0$. g gives the spatial distances between points on an instantaneous space $t = $ constant. In order to guarantee the Euclidean character of these instantaneous spaces, it is assumed that there is a global coordinate system in which $t_{ij} = \text{diag}\,(0, 0, 0, 1)$ and $g^{ij} = \text{diag}\,(1, 1, 1, 0)$.[3]

By an automorphism of L_1 is meant a diffeomorphism d of L_1 onto itself such that $d^*t = t$ and $d^*g = g$. The set A_1 of all automorphisms of L_1 forms a group under composition. In order to derive the structure of A_1, it is convenient to look for the most general coordinate transformation which preserves the canonical forms of the space and time metrics. Leaving aside inversions, which are not of interest here, these transformations are of the form

(1)
$$x'^{\alpha} = R^{\alpha}_{\beta}(t)x^{\beta} + a^{\alpha}(t)$$
$$x'^4 = x^4 + b,$$

where $R^{\alpha}_{\beta}(t)$ is a time dependent orthogonal matrix, $a^{\alpha}(t)$ is an arbitrary smooth function of time, and b is a constant. The automorphisms A_1 are the corresponding point transformations. The structure of A_1 raises three interrelated categories of difficulties for Leibniz.

1. One of the key problems in 17th century mechanics was that of the collision of particles. Is it possible to give an adequate treatment of the laws of collision (which had been derived by Huyghens) within minimal Leibnizian space-time? Ehlers (1973) has pointed out that a primitive treatment is possible under certain assumptions. Suppose that we have N perfectly rigid point particles which change velocities instantaneously and discontinuously upon impact. Then for a collision at $t = 0$, the laws of conservation of momentum and energy can be written in the limiting forms

(2)
$$\lim_{t^- \to 0} \sum_{k=1}^{N} m_k \dot{x}^{\alpha}_k = \lim_{t^+ \to 0} \sum_{k=1}^{N} m_k \dot{x}^{\alpha}_k$$

(3)
$$\lim_{t^- \to 0} \sum_{k=1}^{N} \sum_{\alpha=1}^{3} m_k (\dot{x}^{\alpha}_k)^2 = \lim_{t^+ \to 0} \sum_{k=1}^{N} \sum_{\alpha=1}^{3} m_k (\dot{x}^{\alpha}_k)^2,$$

where $\dot{x} \equiv dx/dt$. It is easy to verify that (2) and (3) are invariant under (1). However, the assumptions underlying (2) and (3) are totally unacceptable to Leibniz. His Principle of Continuity rules out discontinuous velocity changes; and for this reason, Leibniz also rejected perfectly rigid particles. Moreover, Leibniz would not have been happy with such limiting forms of conservation laws since he believed that conservation principles apply to all times.

2. For Leibniz, uniform rectilinear motion has a special status. In addition, he also believed that *vis viva* has an absolute significance. These matters are discussed in detail in Sections 6 and 7 below. But the point to notice here is that there is not enough structure in minimal Leibnizian space-time to support either of these beliefs.

3. A third difficulty was raised by Stein (1975). If the history of a particle is represented by a timelike world line (i.e. a world line which is everywhere oblique to the planes of simultaneity) on L_1, then determinism cannot hold. For among the automorphisms A_1 are those which are the identity on the portion of L_1 on or below some given time slice but which differ from the identity above; such a mapping leaves fixed the entire past history of the particles while changing their future behavior. Since the automorphisms of the space-time time should be symmetries of the dynamical laws (whatever they are), there will be two solutions which describe the same particle histories for all past times but which describe different future behaviors. If one desires determinism, as Leibniz did, then one must either modify L_1 or one must find a different way of describing the history of a particle. Both of these alternatives will be explored below. Ultimately, a combination of the two will be offered as an explication of Leibniz.

2. Enriched Leibnizian Space-times

The straightforward interpretation of uniform rectilinear motion requires the vanishing of the acceleration

(4) $\ddot{x}^\alpha = 0.$

A transformation belonging to (1) will preserve (4) if and only if

(5) $\dot{R}^\alpha_\beta = 0, \qquad \ddot{a}^\alpha = 0.$

The resulting subgroup is linear and, thus, leads to a natural affine structure. This suggests that Leibniz be interpreted in terms of

DEFINITION 2. *Enriched Leibnizian space-time* L_2 is a quadruple (M, t, g, ∇) where M, t, and g are as before and ∇ is a symmetric linear connection such that (i) ∇ is compatible with t and g, i.e. $\nabla t = 0$ and $\nabla g = 0$, and (ii) ∇ is flat, i.e. the Riemann curvature vanishes.

There exists a coordinate system in which

(6) $t_{ij} = \text{diag}(0, 0, 0, 1)$, $g^{ij} = \text{diag}(1, 1, 1, 0)$, $\Gamma^i_{jk} = 0$,

where the Γ^i_{jk} are the components of ∇. The coordinate transformations which preserve (6) (again ignoring inversions) are

(7) $x'^\alpha = R^\alpha_\beta x^\beta + v^\alpha t + c^\alpha$

            $x'^4 = x^4 + b$,

where v^α and c^α are constants and R^α_β is now a constant orthogonal matrix. The automorphisms A_2 of L_2 are the corresponding point transformations.

Although straight line motion and collision problems now make sense, there is still no way to give *vis viva* an absolute significance. This suggests a further enrichment of minimal Leibnizian space-time.

DEFINITION 3. *Doubly enriched Leibnizian space-time* L_3 is a quintuple (M, t, g, ∇, A) where M, t, g, and ∇ are as before and A is a distinguished vector field such that (i) $\nabla A = 0$ and (ii) $t(A, A) = 1$.

There exists a coordinate system in which (6) holds and also

(8) $A^i = (0, 0, 0, 1)$.

The coordinate transformations preserving both (6) and (8) are

(9) $x'^\alpha = R^\alpha_\beta x + c^\alpha$

            $x'^4 = x^4 + b$.

The automorphisms A_3 of L_3 are the corresponding point transformations.

The transformations (7) should seem familiar to the reader. They are, in fact, the Galilean transformations. Indeed, L_2 and L_3 are precisely the most plausible reconstructions of Newtonian space-time, L_2 being Newtonian space-time without 'absolute space' and L_3 being Newtonian space-time with 'absolute space' (the trajectories of the distinguished vector field A being the world lines of the points of 'absolute space').

Leibniz thus seems to be impaled on the horns of a dilemma. On one hand, L_1 is inadequate to support his pronouncements about collisions, straight line motion, determinism, and *vis viva*. On the other hand, adopting L_2 or L_3 makes his view of space-time structure coincide with that of Newton, at least on the phenomenal level.

An easy escape is to seize the second horn and claim that on the phenomenal level, Leibniz had no real quarrel with Newton and that only over the metaphysical foundations do their views clash. But this way out is too easy. It distorts and tends to trivialize one of the most important debates in all of natural philosophy.

3. INTERMEDIATE LEIBNIZIAN SPACE-TIME

In the preceding sections we have detected a tension in Leibniz's writings. Some of his pronouncements drive him towards a spare space-time structure like L_1. But other pronouncements drive him towards richer structures like L_2 or L_3. Can some of the tension be resolved by finding a structure intermediate between L_1 on one hand and L_2 and L_3 on the other? Leibniz's insistence on the relational character of space and motion suggests that we look at the subgroup of the automorphisms A_1 which preserve relative particle quantities, i.e. relative particle distances, relative particle velocities, relative particle accelerations, etc., for arbitrary particle motions. They are the point transformations corresponding to

$$(10) \qquad x'^{\alpha} = R^{\alpha}_{\beta} x^{\beta} + a^{\alpha}(t)$$
$$x'^4 = x^4 + b,$$

where R_β^α is a constant orthogonal matrix. This leads to

DEFINITION 4. *Intermediate Leibnizian space-time* L_4 is a quadruple $(M, t, g, \mathscr{C})$ where M, t, and g are as before and $\mathscr{C}$ is a distinguished atlas of charts such that (i) any chart in $\mathscr{C}$ is canonical with respect to t and g (ii) any two charts in $\mathscr{C}$ are related by (10), and (iii) is maximal with respect to (i) and (ii).

The automorphisms A_4 of L_4 are the isometries of t and g which drag along one chart in $\mathscr{C}$ into another chart in $\mathscr{C}$. They will, of course, be the point transformations corresponding to (10). Indeed, L_4 was constructed in a brute force manner to achieve this result. Once we have L_4, a more intrinsic formulation is forthcoming.

DEFINITION 5. $\mathscr{A} \equiv$ {affine connections ∇: ∇ is flat and $\nabla t = 0$ and $\nabla g = 0$}.

One can see that any ∇ such that $\Gamma_{jk}^i = 0$ for some $\{x^i\} \in \mathscr{C}$ belongs to $\mathscr{A}$. Conversely, if $\nabla \in \mathscr{A}$, then there is a $\{x^i\} \in \mathscr{A}$ such that $\Gamma_{jk}^i = 0$. As a result, the automorphisms A_4 relate the geodesics of the members of $\mathscr{A}$ in that for any $\overset{1}{\nabla}, \overset{2}{\nabla} \in \mathscr{A}$, there is an automorphism which carries the geodesics of $\overset{1}{\nabla}$ onto those of $\overset{2}{\nabla}$; and for every $\overset{1}{\nabla} \in \mathscr{A}$ and every automorphism, there is a $\overset{2}{\nabla} \in \mathscr{A}$ such that the automorphism carries the geodesics of $\overset{1}{\nabla}$ onto those of $\overset{2}{\nabla}$. Thus, another way to think of intermediate Leibnizian space-time is as the quadruple $(M, t, g, \mathscr{A})$.

Note what we do and what we do not have in L_4. We do not have absolute velocity as we do in L_3. Nor do we have absolute spatial acceleration as we do in L_2.[4] We do, however, have absolute spatial rotation. Let V be a unit timelike vector field $(t(V, V) = 1)$ which we may think of as the velocity field of a system of particles. The spatial rotation matrix $\omega_{\alpha\beta}$ of V is given by

(11) $\omega_{\alpha\beta}(V) \equiv V_{\alpha,\beta} - V_{\beta,\alpha}$,

where the covariant components of V are determined by

(12) $V^i = g^{ij}V_j$.

Equation (12) determines V_j only up to a transformation of the form

(13) $V_j \to V_j + \lambda t_j,$ $t_j \equiv \partial t / \partial x^j,$ $\lambda = \text{constant}.$

But since $t_{i,j} = t_{j,i}$, (11) is invariant under (13). Also, for any $\{x^i\}$, $\{x''^i\} \in \mathscr{C}$, the norms $|\omega_{\alpha\beta}|$ and $|\omega'_{\alpha\beta}|$ of the rotation matrices are equal. In the language of $\mathscr{A}$, if V describes a rotation relative to the inertial frames of some member of $\mathscr{A}$, it describes a rotation relative to the inertial frames of all members of $\mathscr{A}$.

If the choice of the standard of inertial motion (the choice of an element of $\mathscr{A}$ or the choice of a reference body which is to be regarded as being 'at rest') is arbitrary, there is no absolute concept of uniform rectilinear motion; but once a choice is made, a standard is fixed. This tack does provide a partial resolution of some of the tension in Leibniz's views. But as we shall shortly see, the resolution is far from complete.

4. LEIBNIZIAN WORLD MODELS

The apparatus developed in the preceding section allows us to offer simultaneously an explication of Leibniz's principle of the Equivalence of Hypotheses about motion and a reconstruction of his theses about the relational character of space and time.

DEFINITION 6. *A Leibnizian pre-model* is a pair (L_4, P) where L_4 is an intermediate Leibnizian space-time and P is the momentum field of a system of particles on L_4 ($P = mV$ where m is the mass and V, the velocity field of the particles, is a unit timelike vector field on L_4).

DEFINITION 7. Two Leibnizian pre-models (L_4, P) and (L'_4, P') are *Leibniz equivalent* iff there is an isomorphism of L_4 onto L'_4 which carries P onto P'.

It is obvious that Leibniz equivalence (which will be denoted by $\equiv$) is in fact an equivalence relation. This makes possible

DEFINITION 8. *A Leibnizian world model* is a $\equiv$-equivalence class of pre-models.

The proposal to be considered now is that we understand Leibniz as saying that physical reality corresponds directly to world models rather than to pre-models.

The defense of this interpretation has two parts. First, it should be evident that two Leibniz equivalent pre-models should count for Leibniz as merely different descriptions of the same physical reality. Leibniz states that it is a mistake to believe that we are describing two different physical situations when we speak of God as having placed bodies in space in one particular arrangement vs. His having translated them one mile to the East while preserving all their internal spatial arrangements.[5] The above definitions make precise the analogue of this idea for the context of space-time, and it is this analogue which Leibniz needs to make good his principle of the Equivalence of Hypotheses about motion.

This leads to the second part of the justification. On the above construal of the Equivalence of Hypotheses, any possible observer or reference body can be regarded as being 'at rest'; for given any timelike world line there is a $\nabla \in \mathscr{A}$ for which the world line is a geodesic, and since ∇ is flat, the equation of geodesic can be expressed in the familiar form $\ddot{x}^{\alpha} = 0$. Moreover, any two possible observers or reference bodies are equivalent for the purpose of describing the relative motion of particles; for given any two timelike world lines, every Leibnizian world model contains two pre-models that are related by an isomorphism of the space-time structures which takes the particle trajectories of one onto those of the other and which takes one of the given world lines onto the other.

It remains to say something about what further conditions Leibniz would impose on the fields P and V of Definition 6. Presumably, his Principle of Plenitude would require him to hold that in a pre-model for the actual world, the velocity field V must be everywhere non-vanishing so that through every point of space-time there passes the world line of some particle. However, on my reading of Leibniz, while the Principle of Plenitude serves as a selection principle for determining the actual world, it need not be satisfied in all possible worlds. This leads us to ask what should be done with the other extreme case where V is everywhere vanishing. On the standard interpretation of Leibniz, such an empty universe model not only does not correspond to the actual world but also may not correspond to any physically possible world. For on the standard interpretation of Leibniz's dictum that 'space denotes, in terms of possibility, an order of things' (Alexander, 1956, p. 26), the 'possibility' is explained by means of relations possible bodies could have to actual bodies (see Bennett, 1974, Ch. 8). When there are no actual reference bodies, this idea collapses. On our interpretation however, the 'in terms of possibility' phrase receives no such simple or direct reading. For example, it is compatible with our interpretation that it is physically impossible for any additional bodies to exist within a range of, say, one hundred miles of any actual body or even for any additional body to exist anywhere. And in terms of *world models* as opposed to *pre-models*, it is not at all unnatural for the Leibnizian to admit the empty universe model as a limiting case.

5. LEIBNIZIAN ALGEBRAS

Can Leibnizian world models be given a direct, intrinsic characterization that does not use in an essential way the very objects – the pre-models – to which Leibniz wants to deny independent reality? And if so, can reasonable laws of motion be formulated directly in terms of the intrinsically characterized world models? Positive answers to these questions would not prove that Leibniz is right. Nor would negative answers refute him. But the questions are urgent. Positive answers would show that Leibniz's sketchy and programatic remarks can be cashed into something more productive than philosophical squabbles. Negative answers would increase suspicions among all but the most dedicated Leibniz devotees that Leibniz was simply playing a game – an amusing game, but one which does not promote a deeper understanding of physics.

Some recent developments in mathematics and physics suggest a way for constructing a positive answer to the first question. Geroch (1972) has noted that in defining the tensor-theoretic apparatus needed for Einstein's general theory of relativity, the underlying differentiable manifold M of space-time is used in only a weak way; namely, it is used to define a ring $\mathcal{R}$ of smooth functions on M and a subring $\mathcal{R}'$ of constant functions. This leads to an approach in which the manifold M

is ignored in favor of a more abstract algebraic structure. An *Einstein algebra* consists of (i) a commutative ring $\mathcal{R}$ (ii) a subring $\mathcal{R}'$ which is isomorphic to the reals and (iii) various tensor-theoretic-type objects and operations defined in terms of $\mathcal{R}$ and $\mathcal{R}'$. (For example, a vector field is construed as a map from $\mathcal{R}$ to $\mathcal{R}$ with certain properties which characterize a derivation.) Of course, Geroch's ideas do not depend in any essential way on relativity theory, and they can be adapted to the Leibnizian context.

Next, recall an important result in topology. Let X denote a compact topological space and $C(X)$ the ring of continuous functions on X. It follows from the work of Gelfand and Kolmogoroff that

$$(14) \qquad X \approx X' \quad \text{iff} \quad C(X) \cong C(X'),$$

where $\approx$ denotes homeomorphism and $\cong$ denotes ring isomorphism.

Together, these two developments suggest a way to complete Leibniz's program. Following Geroch, we can define the notion of a Leibnizian algebra $\mathfrak{A}(\mathcal{L})$ corresponding to a given Leibnizian pre-model $\mathcal{L}$. Then we can try to establish that

$$(15) \qquad \mathcal{L} \equiv \mathcal{L}' \quad \text{iff} \quad \mathfrak{A}(\mathcal{L}) \cong \mathfrak{A}(\mathcal{L}'),$$

This latter relation does indeed hold (but in distinction to (14), no deep mathematics is involved). It might seem, however, that no progress has thereby been made. For have we not simply exchanged one type of equivalence for another, Leibnizian equivalence of space-time models for isomorphism of algebras? As things stand, this is correct. But we can turn our presentation around. Instead of regarding Leibnizian algebras as objects generated by Leibnizian pre-models, we can take the algebras not only as being objects in their own right but also as being ontologically prior to space-time models. The pre-models are now seen as representations or realizations of algebras, where $\mathcal{L}$ represents or realizes $\mathfrak{A}$ if and only if $\mathfrak{A}(\mathcal{L}) \cong \mathfrak{A}$. The equivalence class structure is now explained: $\mathcal{L}$ and $\mathcal{L}'$ are equivalent if and only if they are representations of an algebra $\mathfrak{A}$. In this sense, equivalent space-time models are just different ways of describing the same intrinsic situation as given by an algebra. But why should the algebraic structure be regarded as more basic? These and other problems connected with this proposal will be considered in detail later, and at this juncture I turn to other matters.

In particular, what about our second question concerning the expression of laws directly in terms of the intrinsically characterized Leibnizian world models? There is no *a priori* difficulty here. Einstein's field equations can be expressed directly in terms of Geroch's Einstein algebras. There is no *a priori* reason to think that an analogous thing cannot be done for Leibnizian laws and Leibnizian algebras. But at this juncture, we do not know what Leibniz's laws are, so I pass on to another point.

The original difficulty about determinism raised by Stein disappears. The automorphisms A_4 of intermediate Leibnizian space-time like those of minimal Leibnizian space-time include mappings which are the identity on portions of space-time below a given time slice but are non-trivial above. But in forming Leibnizian world models, we make identifications under the action of the automorphisms A_4. However, it must now be shown how to reformulate the notion of determinism in terms of the intrinsic characterization of world models as Leibnizian algebras. Intuitively, determinism for laws L now means that for any Leibnizian algebras $\mathfrak{A}$ and $\mathfrak{A}'$ obeying L, if the restrictions of $\mathfrak{A}$ and $\mathfrak{A}'$ to a 'thin sandwich' of space-time are isomorphic, then $\mathfrak{A}$ and $\mathfrak{A}'$ are isomorphic. The notion of the restriction of an algebra to a 'thin sandwich' of space-time can be specified entirely in terms of the algebras without mentioning the space-time representations.

6. IMPLICATIONS OF THE EQUIVALENCE OF HYPOTHESES

The above explication of the Equivalence of Hypotheses allows us to evaluate some rather puzzling claims of Leibniz.

In his *Dynamica* (Gerhardt, 1860, pp. 433–514), Leibniz makes a conditional concession to the Newtonians. He concedes that *if* there were solidity of bodies, as it is commonly understood, then the Equivalence of Hypotheses would be broken. Leibniz's reasoning is essentially that employed by Huyghens and then Newton: the tendency of the parts of a solid body to recede from the center (as indicated, for example, by the tension in a cord connecting two globes) would show that the body is rotating; but since the parts of the body are not in relative motion, some additional space-time structure which breaks the Equivalence of Hypotheses is needed to distinguish between cases of rotation and cases of non-rotation.

Leibniz's 'solution' is to deny the existence of solidity of bodies as it is usually conceived. According to Leibniz, what we perceive as solidity is the result of the collective motion of the particles which comprise the body and which act only by contact. This 'solution' will be examined in more detail later. What I want to point out here is that on our interpretation of the Equivalence of Hypotheses, Leibniz did not have to resort to any fancy footwork to handle rotation. For rotation of a system of particles is a well-defined concept in L_4 and is invariant under Leibniz equivalence. And this result is completely independent of whether the cohesion of bodies results from a 'glue', action at a distance, or anything else.

Leibniz might have been somewhat unhappy with this solution since it shows that there is something misleading about the slogan that all motion is the relative motion of bodies. For we do have a kind of motion – rotation – which cannot be analyzed as rotation relative to a system of reference bodies; rather, it is rotation relative to a system $\mathscr{C}$ or $\mathscr{A}$ of abstract space-time structures. Still, these structures are precisely what are needed to make sense of the relative motion of bodies; they arise from the demand that relative quantities be well-defined under the space-time automorphisms and the taking of Leibniz equivalence classes.

Still more puzzling is Leibniz's attempt to reconcile the Equivalence of Hypotheses with an assignment of a subject of motion. In a letter to Huyghens written in 1694, Leibniz stated his view thus:

Even if there were a thousand bodies, I still hold that the phenomena could not provide us (or angels) with an infallible basis for determining the subject or the degree of motion and that each body could be conceived separately as being at rest .... But you will not deny, I think, that each body does truly have a certain degree of motion, or if you wish, of force, in spite of the equivalence of hypotheses about their motions. (Loemker, 1970, p. 418.)

And in the famous fifth letter to Clarke, the distinction between relative and absolute motion is reaffirmed thus:

However, I grant there is a difference between an absolute true motion of a body, and a mere relative change of its situation with respect to another body. For when the immediate cause of the change is in the body, that body is truly in motion.... (Alexander, 1956, p. 74.)

Despite his many protestations, Leibniz appears to be caught in a trap. If the assignment of motive force (*vis viva*) to body *a* rather than to body *b* is to be more than a merely subjective matter, then there must be some underlying space-time structure – like the distinguished vector field of doubly enriched Leibnizian space-time – which makes *vis viva* a well-defined quantity. But such a structure breaks the Equivalence of Hypotheses.

Many commentators have struggled with this dilemma. Some have taken the way out suggested by H. G. Alexander:

Leibniz may therefore have held that the distinction between absolute and relative motion is metaphysical, not physical, that is, the absolute motion of a body can never be experimentally determined; and so the concept of absolute motion is of no use in physics. Such an interpretation is supported by his statement in the *Discourse on Metaphysics*, that moving force is a metaphysical concept. (Alexander, 1956, xxvi–xxvii.)

Alexander's resolution is unacceptable for two reasons. First, by calling motive force a metaphysical concept, Leibniz was not denying it a physical status. In the *Discourse*, the contrast with 'metaphysical' as applied to motive force was not 'physical' but rather 'geometrical'. Leibniz's main goal was to show that there was more to the nature of bodies than the Cartesian 'geometrical' notions of size, figure, and motion (where the quantity of motion is taken as mass times speed). Secondly, in the *Discourse* Leibniz specifically affirms the possibility of determining the subject of motion:

But force or the immediate cause of these changes in the relative situation of bodies is something more real, and there is sufficient basis for ascribing it to one body rather than to another. This, therefore, is also the way to learn to which body the motion preferably belongs. (Loemker, 1970, p. 315.)

Leibniz definitely is in a trap. What we have to try to understand is how he got in it and why he remained ensnared.

In his *Brief Demonstration of a Notable Error of Descartes and Others Concerning a Natural Law* (Loemker, 1970, pp. 296–301), Leibniz sought to out-Descartes Descartes. Leibniz accepted Descartes' axiom that God created a certain amount of motive force which is conserved throughout all time. Then using certain other principles (e.g. conservation of 'elevating force') which he claimed Descartes would have accepted, he shows that motive force cannot be measured by Descartes' quantity of motion (mass times speed) but rather must be measured by mass times speed squared.

It is reasonable to guess that Leibniz took the principle of the conservation of 'elevating force' and the recognition of the importance of *vis viva* directly from Huyghens. (Commenting on Leibniz's *Brief Demonstration*, Huyghens (1686) notes that he had previously (i) recognized the falsity of Descartes laws of impact (ii) formulated a 'truer law' in which the same quantity of motion is conserved in the same direction (conservation of momentum) and (iii) formulated the law of conservation of *vis viva*.) But the concepts borrowed from Huyghens are used by Leibniz in an uncritical way. Huyghens was quite clear about the kinematic relativity of impact phenomena; in fact, he uses this relativity in deriving the laws. Leibniz, however, could not have been clear on this matter at the time of writing his *Brief Demonstration* in 1685 since he takes over Descartes' idea that the total quantity of motive force has an absolute significance and since he asserted that particular bodies can be singled out as possessing motive force while others lack it. Later, it was too late to correct the effects of his original misunderstanding, for his concept of force had become a keystone in his criticism of the Cartesian account of matter and the starting point of his alternative account. In a candid passage in his letter of Huyghens immediately following the passage quoted above, Leibniz confesses:

It is true that from this I draw the conclusion that there is something more in nature than what geometry can determine about it. This is not the least important of the many arguments which I use to prove that besides extension and its variations, which are purely geometrical things, we must recognize something higher, namely, force. (Loemker, 1970, p. 418.)

And, he might have added, it is too late now for me to abandon this conclusion.

7. LEIBNIZ'S ARGUMENTS FOR THE EQUIVALENCE OF HYPOTHESES

When Leibniz speaks of the Equivalence of Hypotheses, he sometimes seems to have something weaker in mind than the dynamical equivalence of all observers formulated above. In the *Dynamica*, written in

the 1690's, Leibniz now emphasizes the kinematical relativity of collision phenomena. Between collisions, the particles travel in uniform
rectilinear motions. Thus, any particle can equally well be regarded as
being at rest. And since indistinguishable causes must give rise to
indistinguishable effects, the laws of collision must respect this
kinematical relativity, or as we would say today, must be Galilean
invariant.

When Leibniz then says that the Equivalence of Hypotheses applies
not only to rectilinear motions but 'universally', he might, as Howard
Stein has pointed out, be referring not to an observer or reference
body but to the interacting bodies themselves (see Stein, 1975). In this
case, the general equivalence of hypotheses would not give the general
relativity of Sections 4 and 5 above, but would simply guarantee that
what holds for impacts also holds for other interactions in which the
trajectories of the particles are curvilinear. Leibniz's claim that all
interactions must ultimately be reducible to impacts is an arguments
that this guarantee is met, and thus can be seen as a mark in favor of
the weaker interpretation of the Equivalence of Hypotheses.

Still, it is evident that Leibniz's doctrines about space, time, and
motion require something akin to our stronger interpretation.
Moreover, Leibniz seemed to have believed that the weaker interpretation provides a basis for the stronger interpretation. He says that all
motions are composed via impacts from uniform rectilinear ones and
that what we perceive as solidity in bodies also results from this
composition (in particular, from the 'crowding together' of the surrounding particles). It is this analysis as applied to circular motion that
Leibniz believes saves the Equivalence of Hypotheses in the strong
sense.

From these considerations it can be understood why I cannot support some of the
philosophical opinions of certain great mathematicians on this matter, who admit empty
space and seem not to shrink from the theory of attraction but also hold motion to be an
absolute thing and claim to prove this from rotation and the centrifugal force arising
from it. But since rotation arises only from the composition of rectilinear motions, it
follows that if the equipollence of hypotheses is saved in rectilinear motions, however
they are assumed, it will also be saved in curvilinear motions (Loemker, 1970, pp.
449–450).

Leibniz's arguments simply do not mesh. If we have the full dynamical equivalence of hypotheses, any possible observer or reference body

can be regarded as being at rest. But this general dynamical relativity destroys the uniform rectilinear behavior of bodies in impact phenomena. Put in terms of our technical apparatus, we cannot expect that the element ∇ of $\mathscr{A}$ which has as a geodesic the trajectory of the arbitrarily chosen observer or reference body will also have among its geodesics the trajectories of particles interacting via impacts. Nor can we expect the laws of conservation of momentum and energy to hold if the velocities of the particles are measured relative to our arbitrary observer. Conversely, the postulation of a special ∇ which will make possible the validity of the laws of impact in their usual form will destroy the dynamical equivalence of hypotheses.

Leibniz's assertion that all interactions are reducible to impacts and that all motions are composed, via impacts, of uniform rectilinear ones was, in effect, the announcement of a research program. Can the program be carried out for the interactions studied in the 17th century? Gravitation provides the main test case. Since on Leibniz's assumption bodies act only by contact and move with uniform rectilinear motions between impacts, it is amusing to ask whether the motions of the planets as predicted by Newton's theory can be regarded as the geodesics of an affine connection ∇_N. The answer is in the affirmative (see Earman and Friedman, 1973). But ∇_N is curved, a possibility which Leibniz would have had no inkling of how to describe. And the use of ∇_N implies in a dramatic way that space-time has a causal effect on material bodies, a result repugnant to Leibniz. Nor is there any way to broaden the class $\mathscr{A}$ so as to include curved connections since then the mappings which carry the geodesics of one member onto the geodesics of another will not preserve the relative motion of bodies.

The other way to pursue Leibniz's research program is to treat the trajectories of the planets as resulting from effectively continuous collisions from some underlying subtle medium. This, of course, is the route Leibniz took. In his *Testamen de Motuum Coelestium Causis* (Gerhardt, 1960, pp. 144–187) and subsequent works, he attempted to construct a vortex theory of planetary motion. All of these attempts were fraught with major difficulties (see Eiton, 1972). Still, the general research program was laudable, and it did produce gains in understanding.

8. A REASSESSMENT OF LEIBNIZ'S VIEWS ON SPACE AND TIME

The conclusion that intermediate Leibnizian space-time and Leibnizian world models do not provide a viable basis for reasonable laws of motion does not entirely destroy the thrust of Leibniz's conception of space and time. For to a certain extent, Leibniz's views on the foundations of space-time are separable from his views on motion. Suppose, for sake of argument, that Leibniz had agreed that something on the order of enriched Leibnizian space-time L_2 (or Newtonian space-time without absolute space) is needed for physics. Still, he would have insisted that world models be formed by taking equivalence classes under the action of the automorphisms A_2 of L_2. An intrinsic characterization of these new world models can, of course, be given along the lines suggested in Section 5. The result can be called *Leibniz-Newton algebras*. Leibniz's doctrine would then be expressed by the assertion that two or more representations of a given Leibniz-Newton algebra are only different descriptions of the same physical reality.

Leibniz would defend his procedure by asking why God should actualize one representation rather than another, and would conclude that since He could have no sufficient reason for choosing, the choice must be chimerical. But the focus of the issues here is not so much God as Man. For example, we mortals want to make predictions, and we want to be able to confirm or disconfirm theories by observational checks on predictions. How these earthly desires can be satisfied in terms of algebras without making use of their space-time representations remains to be seen (see Geroch (1972) for some suggestions).

The main virtue of the Leibniz and the Leibniz-Newton algebras – their representation free character – also raises another problem. In addition to the intended realizations in terms of the classical space-times L_4 and L_2 respectively, there are other realizations which do not correspond to any ordinary space-time model. What should be done about such non-standard realizations? And what do they show about our ordinary conceptions of space and time. It might be thought that Leibniz's monadology has some relevance here. His mature monadology is couched in terms of 'perceptions' of the monads, and since perception terminology suggests spatio-temporal concepts, it might be hoped that the machinery of the monads would help to select out the space-time realizations. It is unlikely, however, that this hope can be fulfilled. For although Leibniz used perception terminology because of its rich connotations, his technical explanations of the meaning of 'perception' use such very general notions – e.g. correspondence and isomorphism – that it is unlikely that 'perception' can play the envisioned selection role.

On the other hand, the monads may serve another need. The Leibniz (or the Leibniz-Newton) algebras can at best be seen as providing only a preliminary solution to the problems which Leibniz posed for Newton's doctrines of space and time. The reason, of course, is that there are distinct but isomorphic algebras. Since God could presumably have no reason to choose one of the isomorphic algebras over another, the choice must be a chimerical one, and there must be an even deeper level of reality containing elements for which the isomorphic algebras are different representations. If this argument is not to iterate over and over again, forcing us ever further down into an unending sequence of levels of reality, we must eventually arrive at an uhr level in which distinct but isomorphic situations are not permitted. Perhaps the monadology can be taken to describe this uhr level. Unfortunately, this would appear to be incompatible with the phenomenalistic interpretation of the monadology towards which I believe Leibniz was tending in his later years.

In any case, the issues I have drawn from Leibniz's writing transcend both the historical setting and the peculiarities of Leibnizian metaphysics. These issues apply just as urgently to 20th century theories of space-time as to 17th century theories.

Department of Philosophy,
University of Minnesota

NOTES

* This paper was stimulated by lectures and papers by Howard Stein; but this does not imply that he would agree with either my approach or my conclusions. I am indebted to Paul Fitzgerald and other members of the audience at the Fifth International Congress of LMPS for several suggestions. I also wish to thank Professor Lawrence Markus for helpful conversations.

[1] Not everyone agrees with this reading of Leibniz's monadology; for a contrary opinion, see Ishiguro (1972).

[2] What further light the monadology casts on the nature of spatio-temporal relations is a difficult and controversial question; see Section 8 below for some remarks on one aspect of this question.

[3] For further details, see Earman and Friedman (1973). Latin indices run from 1 to 4 while Greek indices run from 1 to 3. The Einstein summation convention on repeated indices is used throughout.

[4] In L_2 the instantaneous acceleration vector $a^i = d^2x^i/dt^2 + \Gamma^i_{jk}(dx^j/dt)(dx^k/dt)$ is always tangent to a plane of absolute simultaneity and, thus, its spatial length is well-defined.

[5] Actually, the example Leibniz uses in this third letter to Clarke (see Alexander, 1956, pp. 25–30) is couched in terms of changing 'East to West'. This suggests the operation of mirror image reflection which leads to problems about incongruous counterparts; these problems will be ignored here.

BIBLIOGRAPHY

Alexander, H. G.: 1956, *The Leibniz-Clarke Correspondence*, University of Manchester Press.

Bennett, J.: 1974, *Kant's Dialectic*, Cambridge University Press.

Earman, J. and Friedman, M.: 1973, 'The Meaning and Status of Newton's Law of Inertia and the Nature of Gravitational Forces', *Philosophy of Science* **40**, 329–359.

Ehlers, J.: 1973, 'Survey of General Relativity Theory', in W. Israel (ed.), *Relativity, Thermodynamics, and Cosmology*, D. Reidel, Dordrecht-Holland.

Eiton, E. J.: 1972, *The Vortex Theory of Planetary Motions*, American Elsevier, New York.

Gerhardt, C. I.: 1860, *Leibnizens mathematische Schriften*, H. W. Schmidt, Halle, Vol. 6.

Geroch, R.: 1972, 'Einstein Algebras', *Journal of Mathematical Physics* **26**, 271–275.

Huyghens, C.: 1686, 'Considerations sur la conservation du mouvement ou du force', *Oeuvres Completes de Christiaan Huyghens*, M. Nijhoff, 1937, Vol. 19, pp. 162–165.

Ishiguro, H.: 1972, *Leibniz's Philosophy of Logic and Language*, Cornell University Press, Ithaca, N.Y.

Loemker, L. E.: 1970, *Gottfried Wilhelm Leibniz, Philosophical Papers and Letters*, D. Reidel, Dordrecht-Holland.

Stein, H.: 1975, 'Some Philosophical Pre-History of General Relativity', forthcoming in *Minnesota Studies in the Philosophy of Science*, **8**.

JÜRGEN MITTELSTRASS

CHANGING CONCEPTS OF THE *A PRIORI*

I

The history of philosophy and science can be viewed from several perspectives. From an external point of view, concerning the so-called socio-economic conditions of theory construction, and from an internal point of view which concerns the logical structure of theory. One important internal point of view involves the effort to give a foundation to theoretical propositions which can be established by itself without recourse to any other propositions, especially not empirical ones. A proposition that possesses such a foundation, is normally called *a priori*. The question is whether there are *a priori* propositions, i.e. whether propositions in theoretical contexts can be so identified that they fulfill the cited condition of independence. In regard to the history of philosophy and science it seems as if this is still an open question; one often hears .it said that the problem is as old as philosophy itself (a formulation which usually disguises an acute theoretical weakness). Those who answer affirmatively are called rationalists; those who answer negatively empiricists.

This classification is admittedly not only rather crude, it already presupposes a particular formulation of the distinction between *a priori* and non-*a priori* propositions, namely the identification of non-*a priori* with 'empirical'. In fact, current usage of '*a priori*' is as a rule such that the expressions '*a priori*' and 'empirical' are regarded as mutually exclusive. At the same time it is asked whether there are *a priori* elements in empirical knowledge. This question is preferably considered (as already with Kant) in reference to physics. Again we understand by *a priori* elements a proposition whose validity (truth) is independent of the validity (truth) of physical theories with regard to their empirical portions in the strict sense. The thesis of the 'apriorists' among the epistemologists is that the converse case, namely that there is empirical knowledge (such as physics) whose validity is completely independent of *a priori* knowledge, is not conceivable.

Butts and Hintikka (eds.), Historical and Philosophical Dimensions of Logic, Methodology and Philosophy of Science, 113–128.

Again, we can introduce in a highly simplified form three different views on the question of the possibility of *a priori* foundations: (1) There is no *a priori* knowledge. All knowledge is entirely empirical, i.e. empirically based. This position could be called extreme empiricism. (2) There is *a priori* knowledge, but such knowledge plays no role in determining the foundations of empirical knowledge. This position could be called moderate empiricism. (3) There is *a priori* knowledge and it does play an essential role in providing the foundations of empirical knowledge. This position, with reference to the traditional historical classifications, could be called moderate rationalism. In addition, there is finally the Leibniz thesis (one could just as well call it the Descartes thesis), that all knowledge, including empirical propositions which occur in the empirical sciences, can be reduced to (or analyzed in terms of) *a priori* propositions. Since this thesis has played no role in the epistemological discussions of the foundations of the empirical sciences (at least not since Kant), I am content just to mention it. As a specific position it could be called extreme rationalism.

The identification of non-*a priori* with empirical and the attendent diametrical contrast between *a priori* and empirical (on which these classifications rest) is by no means self-evident. Such a restriction of the earlier uses of '*a priori*' has only been common since Kant. This restriction makes it very difficult to recognize the distinctions which have to be made, if we are to bring to a decisive conclusion the long-standing controversy about the concept of *a priori* truth. The difficulties here arise above all in connection with the use of the terms 'empirical' and 'experience' (and this was already the case for Kant). The reasons for this must be carefully analyzed. My aim is therefore two-fold. It is (1) to determine the various concepts of the *a priori* in their historical development, and (2) to point out some distinctions, which will provide conceptual clarification of these difficulties. My aim is therefore simultaneously historical and systematic.

II

First let us turn to the usage of the term '*a priori*' before Kant.[1] This usage can be characterized by saying that the opposition between

empirical and non-empirical knowledge, as mentioned above, originally played no role. The opposition *a priori – a posteriori*, which has
gradually emerged since the twelfth century within the context of the
distinction between arguments *ex prioribus* and arguments *ex post-
erioribus*, has strictly speaking a *proof-theoretical* meaning, which has
no connection with an identification of *a posteriori* and empirical which
is very characteristic of Kant's usage. In the fourteenth century, for
instance, Albert of Saxony finally stabilized this proof-theoretical
usage, which derives, among others, from Boethius, Abelard, and
Thomas Aquinas. According to this tradition *a priori* arguments serve
to identify a proof which proceeds from causes to effects (or, in a
logical mode, from premisses to conclusions). In the language of the
time: *demonstratio procedens ex causis ad effectum*. By contrast, *a
posteriori* arguments serve to identify proofs which derive causes from
effects (or premisses from conclusions). The technical term for this
was: *demonstratio procedens ab effectibus ad causas*. In the first case it
is, according to Albert of Saxony, a matter of *demonstratio a priori*; in
the second case, a matter of *demonstratio a posteriori*.[2]

By means of the concepts of *a priori* and *a posteriori*, the medieval
philosophers reconstruct the Aristotelian distinction between 'prior
according to nature' (πρότερον φύσει) and 'prior according to knowledge' (πρότερον πρὸς ἡμᾶς)[3], which recognizes the fact that, as a rule,
effects and conclusions (but not their causes and premisses) are immediately known. The proofs *a priori* are related to propositions which
are 'prior according to nature', and the proofs *a posteriori* are related
to propositions which are 'prior according to knowledge'.[4] In view of
these historical connections it is thus comprehensible that this reconstruction of an older distinction was immediately linked to another
Aristotelian distinction, which also belongs to a proof-theoretical context. I refer to the distinction between a proof which explains what is in
terms of why (διότι) it is (*demonstratio propter quid*), and the demonstration that (ὅτι) something is (*demonstratio quia*). Already with
Thomas Aquinas (if not even earlier) there comes to be an explicit
connection between *demonstratio a priori* and *demonstratio propter quid*
(*cur ita sit*), on the one hand, and between *demonstratio a posteriori*
and *demonstratio quia* (*quod ita sit*), on the other hand.[5] Finally Albert
of Saxony transforms this connection into a virtual identification.[6] That

proof-theoretical viewpoints remain predominant is made clear by the difficulties which arise in connection with *indirect proofs*. William of Ockham classifies these proofs (which neither provide premisses in a strict sense nor argue from supposedly true conclusions) as a special type of *demonstrationes quia*, which are *a priori* in contrast to the usual subordination of *demonstrationes quia* under *demonstrationes a posteriori*.[7] Later, indirect proofs will, for the same reasons, be regarded as proofs *a posteriori*.[8]

It should be clear that such conflicting classifications as we see here in connection with indirect proofs, are generally only intelligible if the contrast *a priori – a posteriori* only concerns itself with the question whether the proof of a proposition must depend upon conclusions which one has drawn from it (*demonstratio a posteriori*), or if such a dependence does not exist (*demonstratio a priori*). A reduction of *demonstratio a posteriori* to empirical connections does not take place, not even in connection with *demonstratio quia*. This very circumstance illustrates the proof-theoretical nature of the distinction *a priori – a posteriori*, which is characteristic of the original meaning of the distinction. In view of this connection we should speak here of a *proof-theoretical a priori*.

Obviously, this proof-theoretical *a priori* cannot be blamed for the ensuing controversies about the possibility of *a priori* knowledge. In addition, these controversies depend upon the later identification of non-*a priori* and empirical. This identification is made possible by the discussions on method, which accompanied the origins of modern physics. Characteristic likewise for these discussions is the appropriation of a distinction, originally Aristotle's, namely, between *analytic* and *synthetic* methods and the related distinction between deduction and induction. What was regarded by Aristotle and the medieval philosophers (especially in their discussions of the *Posterior Analytics*) as a proof-theoretical matter comes to be regarded by the Paduan Aristotelians (e.g. Zabarella) in the form of the distinction between the *metodo risolutivo* and the *metodo compositivo* as the cornerstone of a methodology of the empirical sciences.[9] So, too, for Galileo, who takes this distinction over from the Aristotelians.[10] For him, the *metodo risolutivo* serves as an instrument for discovering propositions with which to explain observed phenomena, while the *metodo compositivo* leads with the help of these propositions to the formulation of hypotheses, which are confirmed by further application of the *metodo risolutivo*. Galileo, in doing this, draws criticism from Descartes, who, using the older terminology, insists that Galileo has only explained '*quod ita sit*', whereas he himself has thought '*cur ita sit*'.[11] Newton finally shifts this distinction (which, even with Galileo and Descartes, remains proof-theoretical) decisively in the direction of an empiricist methodology of cause and effect: the analytic method has the task of finding causes for observed effects, while the synthetic method argues from observed causes to effects.[12] In other words, in contrast to Galileo's procedure, in which the *metodo risolutivo* is still regarded an *a priori* clarification of the basic concepts and propositions, Newton now subjects analysis also to empirical controls. To the extent that analysis becomes part of a total empirical procedure, the identification of non-*a priori* and empirical is coming to fruition.

The first terminological specification takes place at the hands of Leibniz who, on the one hand, calls the method which establishes reasons (*rationes*) for empirical connections a method *a priori* (*a priori per rationes*)[13], but, on the other hand, insists that this cannot be achieved without supplementation by '*experimento a posteriori*', i.e. by experience.[14] Leibniz, too, utilizes a proof-theoretical sense of the *a priori* and even seeks to give it within the context of his analytical theory of concepts and propositions a methodologically well-founded and precise formulation. It should be noted, however, that in this connection he modifies totally the conceptual framework by regarding a proof *a priori* (*probatio a priori*) as '*independens ab experimento*'.[15] The claim that even contingent (empirical) propositions, which initially are known only *a posteriori*[16], can in principle be proven *a priori*[17] (at least for God), underscores this: whether a proposition can 'only' be known *a posteriori* or can be proven *a priori* depends upon how much *a priori* knowledge we already possess. Small wonder, that for Christian Wolff finally every proof, which requires intermediate steps (*ex aliis cognitis ratiocinando*)[18], represents a proof *a priori* and that *demonstratio a posteriori* refers only to immediate (sensual) experience.

This modification which took place hand in hand with the emergence of modern physics should be characterized as a transition from a proof-theoretical *a priori* to a *foundational a priori*. If one is speaking in the seventeenth and eighteenth centuries of the possibility of *a priori* knowledge, this is equivalent to speaking primarily about the foundational problems of the empirical sciences, particularly physics. That is true not only for 'rationalists' such as Descartes and Leibniz, but also for 'empiricists' such as Newton and Locke. Implicitly, the proof-theoretical contrast *a priori* – *a posteriori* becomes transformed into the mutually exclusive opposition between *a priori* and empirical, generating numerous confusions later on.

It was Kant who brought this modification explicitly into the foreground and who at the same time gave the definite impetus to all further discussions up to and including contemporary controversies. This takes place in the *Kritik der reinen Vernunft* with the help of the two contrasts analytic – synthetic and *a priori* – *a posteriori* (i.e. *a priori* – empirical), which represent apart from the connection analytical – empirical two distinctions completely independent of one another. The aim of a proof-theoretical endeavour, understood in the foregoing sense, is to determine the meaning of the connections analytic-*a priori*, synthetic – *a priori*, and synthetic – *a posteriori*; the connection analytic – *a posteriori* is excluded because of the definition of 'analytic' as an inclusive relation of concepts (the converse a-relation between concepts in propositions with a simple subject-predicate structure).[19] Analytic propositions turn out on this analysis to be readily demonstrable as *a priori*. But this already anticipates a foundational connotation for '*a priori*'. Kant calls propositions '*a priori*' if they give "von der Erfahrung und selbst von allen Eindrücken der Sinne unabhängiges Erkenntnis";[20] and he calls them 'empirical' if they have "ihre Quellen *a posteriori*, nämlich in der Erfahrung".[21] But if all analytic propositions are necessarily propositions *a priori*, then all non-*a priori* propositions are necessarily synthetic, namely non-analytic ("Erfahrungsurteile, als solche, sind insgesamt synthetisch").[22] As a result of this, the central question remaining for Kant is to determine the meaning of the connection 'synthetic – *a priori*'. As he puts it: determining the possibility of synthetic propositions *a priori*. Such propositions are synthetic, i.e. not fully determinable by analytic, conceptual explication, but notwithstanding this, they have an *a priori* validity, i.e. they are not (despite their synthetic character) refutable in terms of the conclusions which follow from them (in Kant's own words: they are not refutable by experience).

The answer to this question, as is well known, is offered by an arithmetical example, which shows that we are dealing here with propositions whose validity is not only based upon logical and definitional (analytical) factors but is based rather upon *constructions*, which themselves are not based entirely upon logical and definitional determinations of the utilized concepts. Besides geometry and arithmetic there are, above all, definite foundational principles of physics (for instance, the so-called causal principle), whose validity is claimed to be derived from their synthetic *a priori* character. Thus, in contrast to Leibniz and also to later formulations within Logical Empiricism, a distinction between analytic and empirical is, viewed from a methodological prospective, incomplete.

My aim here is neither to give a full account of Kant's reflections on these distinctions and questions, nor to judge whether he has succeeded in explaining adequately his concept of the synthetic *a priori*. My aim, rather, is only that of determining the meaning of the contrast *a priori – a posteriori*, which arises from the identification of '*a posteriori*' and 'empirical'. In the sense relevant here Kant means by 'empirical' an *empirical-physical* proposition, i.e. a proposition which in contrast to 'experience' in a pre-theoretical (namely ordinary life) sense is already based upon *results of measurement*. In this sense Kant also speaks of "empirische Erkenntnis".[23] The distinction between the empirical validity and *a priori* validity of propositions has, therefore, to be understood in such a way that the empirical validity rests upon the results of measurement (especially measurements of length, time and mass), whereas *a priori* validity arises when the truth of any proposition does not rest upon such measurements.

A methodological connection between empirical validity and *a priori* validity is constructed by Kant utilizing the claim (and its justification) that "empirische Erkenntnis" (empirical knowledge) cannot be properly established independently of *a priori* knowledge. And it is precisely here that the concept of the *a priori* acquires its *foundational* aspect with reference to the methodological construction of the empirical sciences. What chiefly matters for Kant in his application of the contrast *a priori – a posteriori* is to show that the empirical sciences such as physics in both their methods of measurement and their leading orientations (e.g. basic principles such as the causal principle) are dependent upon non-empirical propositions, whose validity is entirely *a priori*. This emerges clearly from Kant's manner of speaking about the conditions for the possibility of experience as well as from the fact that '*a priori*' usually is intended by Kant to refer to the '*a priori* part of empirical-physical knowledge' (e.g. what he means by 'experience'). In his foundational formulation of the *a priori* Kant thus belongs to the moderate rationalists mentioned above.

In following Kant without thereby committing ourselves to defending his position in detail, the following more precise definitions can be suggested: (1) a proposition *a* is true *a priori* (or valid *a priori*), if its proof can be given without reference to the consequences which follow from *a*. This definition, which rests upon a synonymy between 'true' and 'proven', can still be understood *proof-theoretically*, insofar as it makes no reference to the kinds of consequences which follow from *a*. (2) Then is introduced the *foundational* restriction, which has been responsible for most of the subsequent epistemological controversies about the status of *a priori* proofs: a proposition a_1 is true *a priori* (or valid *a priori*) only in relation to another proposition a_2 which results from certain definite measurements (in other words, a_2 is an empirical-physical proposition). Logically viewed, the '*a priori*' is a dyadic relation between propositions (or propositional systems), one of which must be empirical.

It is clear that all *analytic* propositions constitute a part of *a priori* knowledge, whether they are founded upon logico-definitional conventions (i.e. *material-analytic* propositions) or whether they are true only in virtue of their logical form (i.e. *formal-analytic* propositions). And this is independent of Kant's concept of the analytic (or analytic judgements), which refers back to Leibniz.

III

With respect to the question whether *a priori* truth is identical with analytic truth (which was answered positively by Leibniz and later by the Logical Empiricists) Kant's formulation of the principles which constitute rational physics is not the only meaningful answer; indeed, it remains highly controversial. Such considerations as one sees within phenomenological epistemology, especially Husserl's, make this clear. They are directed against, for instance, the psychologistic residue exhibited by Kant's epistemological conceptions, insofar as certain formulations seem to entail that the synthetic *a priori* is concerned with the internal structure of consciousness. In German Idealism (Fichte, Schelling), but also in an especially extreme form with Helmholtz, this *consciousness-theoretical* approach reached its final form, Kant's foundational concerns to the contrary notwithstanding.

According to Husserl our interaction with the world, even if this interaction finds expression in theory, takes the form of an "empirischer Gesamtstil".[24] In the first instance, this is a matter of daily activities, on which ultimately are based theoretical interests and their realization within theories. Husserl has thereby drawn attention to a distinction between two concepts of experience, which were in danger of being ignored by virtue of the polarization between the *a priori* and the empirical (and thereby with the related interpretation of 'empirical' as 'empirical-physical'). His claim that measure theory is based in pre-theoretical activities[25] further elaborates Kant's reflections on the *a priori* character of geometry and arithmetic (pragmatic considerations thereby replacing preoccupations with consciousness). This can be viewed as a clear anticipation of an operational theory of geometry, which today is being elaborated within the school of constructivist epistemology.[26] With regard to the foregoing foundational conclusions the pre-theoretical activities assume in Husserl the character of an '*lebensweltliches Apriori*'.

If one ignores Husserl's effort to find a 'transcendental' justification for the '*lebensweltliches Apriori*' (thereby compromising his own deepest insights) one can characterize his articulation of Kant's analysis as the attempt to re-introduce into the contemporary concept of experience an older Aristotelian notion of experience. This might seem to run counter to the actual evolutionary origin of contemporary science. But all that is meant by '*Lebenswelt*' in Husserl's sense, is nothing other than what Aristotle meant by 'experience', namely the capacity of safe orientations, learned in concrete situations of activity. This could also be referred to as the familiarity with and the control over everyday activities and occurrences without recourse to any previously acquired and theoretically articulated knowledge. On the other hand, it is the gist of the foregoing distinction that theory construction is dependent upon this elementary knowledge and capacity; in other words, 'experience', understood, of course, in a non-Kantian sense, acquires an *a priori* dimension.

Because not all portions of a general pre-theoretical *Lebenspraxis* belong to the *a priori* elements of empirical theories such as physics or to non-empirical theories such as arithmetic, this issue naturally requires a more *systematic* formulation. Because of limitations of space and because these matters are not directly *historical*, I will content myself here with a very brief treatment of them.[27]

The elementary knowledge and activity, on which depends the methodological construction of theoretical knowledge, is given, in the first place, through the learning of, and the control over, *distinctions* and *orientations*. Through distinctions (the action of distinguishing) and orientations (the action of orientating) is constituted a pre-theoretical praxis, which can be characterized as an integrated *teaching- and learning-situation*. In such a situation simple speech-acts (such as making distinctions) and non-linguistic activities (such as visualizing objects in space) are *pragmatically* controlled. The subordination of such practical activities to *norms* (e.g. of a factual praxis of using distinctions, however complex it might be, to an introductory praxis) leads to a reconstruction of the distinctions-knowledge given with this praxis, including the reconstruction of elementary proofs, such as the 'argumentative' establishment of elementary propositions, by means of which the first simple questions of validity can already be answered. Conversely, every theoretical proposition depends upon this elementary praxis of distinguishing, of using distinctions and of arguing for and against claims dependent upon distinctions. There is no 'beginning' in the strict sense for a rational discourse, which could stand independently of distinctions, and thus no 'beginning' for theoretical discourse, which could dispense with a normative formulation of elementary speech-acts and with those elementary acts of inference associated with them. The same is true, of course, for a definitional or axiomatic construction of knowledge.

In other words, theoretical relations are based upon an elementary knowledge of distinctions and orientations. To formulate the point methodologically, they are based upon the knowledge given by elementary predication, i.e. by elementary propositions and their logical nexus, corresponding to elementary predication. To the genetic primacy of a knowledge of distinctions and orientations being available within the *Lebenswelt* there is associated the logical primacy of the elementary predication with respect to complex propositions which build up theories and to specialized proof methods which are applied here. Since the theoretical knowledge does not allow of either explanation or proof without making use of elementary predication and, further, since no *a priori* proof of a theory can be given, in which we do not already find, as *a priori* elements, certain pre-theoretical activities of distinction and orientation, as a result elementary predication, epistemologically conceived, is an *a priori* part of all knowledge. I want to call this *a priori*, which is part of the *lebensweltliches Apriori* in Husserl's sense, the *distinction-a priori* (*Unterscheidungs-Apriori*).

If one recognizes such knowledge as is given with the elementary predication and its corresponding elementary propositions as experience in its elementary (Aristotelian) sense, then such experience, of course, does not involve the opposition, but rather the element of the distinction-*a priori*. Most of the difficulties which are connected with the concept of *a priori* knowledge arise as a result of the fact that no distinction is made between the concept of a pre-theoretical experience and the concept of a technologically feasible experience, i.e. scientific experience in its strict sense, and elementary knowledge of distinctions and orientations was (mis)understood as a partial empirical prerequisite of scientific experience. Only so conceived, could the impression arise, that an *a priori* foundation of empirical theories could generally only be possible at the expense of the empirical content of such theories.

Moreover, the reference to a distinction-*a priori* is by itself still unsufficient for comprehending the structure particularly of empirical theories on an adequate methodological foundation. This would be to overlook the fact that an empirical theory, such as one finds in physics, does not consist only of linguistic norms, the foundations of which can be, in principle, controlled by a praxis of argumentation. Presupposed by such a theory is, furthermore, a special technical knowledge (*Herstellungswissen*) which is utilized in the experimental structure of empirical theories. This technical knowledge is not, at least not as a rule, a matter of argument. It shares with argumentative knowledge, constructed on the basis of an elementary praxis of distinctions and orientations, the property that it is not yet theoretical knowledge. Such knowledge is, moreover, such that it emerges (in a *technical* praxis) from elementary technical activities, which are prior to any *theoretical* control over such activities. Examples of this can be found in the manufacture and utilization of such simple tools as the hammer and wedge, the building of natural materials such as wood and clay, as well as construction in its rudimentary forms, and as stacking things on top of one another. These examples may seem simple-minded, but the fact is that all technological praxis (i.e. the construction and use of complicated machines, even in their modern and therefore theoretically-dependent form) still rest upon these elementary capacities and the knowledge represented by them. This is even true with regard to those causal propositions central for a technological praxis, insofar as only someone who has himself learned in concrete activity to recognize the consequences of activity and those things which depend on it, can know what a 'causal connection' is.

In other words, the methodological structure of empirical knowledge, given in the form of an empirical theory, depends not only upon a pre-theoretical knowledge of distinctions and orientations but also upon a pre-theoretical technical knowledge, in this case the technically successful praxis of manufacturing (simple) tools and instruments. On analogy with the earlier distinction-*a priori*, I want to call this *a priori*, which is also part of the *lebensweltliches Apriori*, the *technical-a priori* (*Herstellungs-Apriori*). In case of physics this technical-*a priori* can be developed into a *measuring-a priori* (*meßtheoretisches Apriori*) via a theory of length-, time-, and mass-measurement, which bears the same foundational relation to an elementary praxis of distinctions and manufacturing (thus being, in this sense, also not axiomatic).

In geometry, which is thereby determined in Kant's sense as an *a priori* part of physics, such a foundational relation is built up by defining, in the framework of an *operational* theory, the geometrical concepts in terms of manufacturing procedures of spatial forms. Conceived of as a theory of spatial forms and magnitudes, geometry itself is traced back to a technical praxis, which is wholly independent of geometrical determinations (determinations which already presuppose the manufacture of measuring instruments). The actual, hypothetico-deductive structure of an empirical theory like physics is based upon the measuring-*a priori*, the latter being founded in a pre-theoretical knowledge and ability. Proceeding from a modified and complemented conception of *a priori* elements within even empirical theories, this structure can be described in three steps:

(1) The basis of any theoretical knowledge is a pre-theoretical *a priori* given, on the one hand, by an elementary knowledge of distinctions and orientations, and, on the other hand, by an elementary technical knowledge. Theoretical orientations remain dependent upon this pre-theoretical *a priori*.

(2) On the basis of a pre-theoretical *a priori* is built up a measuring-*a priori*. By resort particularly to the elementary technical knowledge and the norms guiding this knowledge, the measuring-*a priori* sets the standards for the procedures of measurement which guarantee the objectivity of a measuring (empirical) theory.

(3) Based upon a theory of measurement, founded in such a way, there follows the construction of an hypothetico-deductive empirical theory, which depends upon procedures of measurement.

We are dealing here with the articulation of a *graded a priori* which (in contrast to many other customary concepts of *a priori*) is not restricted to *analytical*, i.e. logico-definitional, relations as well as the Kantian expansion of these relations into so-called *transcendental conditions* of measuring theories. If one calls a theory which reconstructs an elementary praxis of distinctions an (elementary) theory of predication, and if one, furthermore, calls a theory which reconstructs an elementary technical praxis (according to the Aristotelian distinction between praxis and *poiesis*) a theory of *poiesis*, then those relations which are dealt with within a theory of predication and a theory of *poiesis* belong to the so-called *a priori* foundation of theories, whether empirical or non-empirical. Only then can the *foundational* aspect of the concept of *a priori*, which since Kant has thrust the original *proof-theoretical* aspect into the background, gain its full epistemological significance.

University of Constance

ACKNOWLEDGEMENT

I would like to thank Laurens Laudan (University of Pittsburgh) for his help in preparing the English text.

NOTES

[1] A good survey of the relevant material can be found in Schepers (1971).

[2] Cf. Prantl (1855–70, 4, p. 78). The distinctions come from Albert of Saxony's commentary on the *Analytica Posteriora*.

[3] *An. post.* A2.71b33–72a5.

[4] The proof-theoretical character of this distinction is emphasized by the determination of the logical fallacy of *inversion* (with the help of the concept ὕστερον/πρότερον): this fallacy exists if the proof of a proposition B from the premiss A makes use not of the implication A < B (A implies B) but, incorrectly, by the converse implication B < A (B implies A). Cf. Aristotle, *An. pr.* B16.64b28–33.

[5] "[· ·] *duplex est demonstratio. Una quae est per causam, et dicitur propter quid: et haec est per priora simpliciter. Alia est per effectum, et dicitur demonstratio quia: et haec est per ea quae sunt priora quoad nos. Cum enim effectus aliquis est nobis manifestior quam sua causa, per effectum procedimus ad cognitionem causae*" (*Sum. Theol.* I, qu. 2, art. 2c).

[6] "*Demonstratio quaedam est procedens ex causis ad effectum et vocatur demonstratio a priori et demonstratio propter quid et potissima; [· · ·] alia est demonstratio procedens ab effectibus ad causas, et talis vocatur demonstratio a posteriori et demonstratio quia et demonstratio non potissima*" (*Quaestiones subtilissimae Alberti de saxonia super libros posteriorum*, Venice, 1497; quoted from Prantl (1855–70, 4, p. 78)).

[7] *Summa Totius Logicae* III, IIc. 17/19; quoted in Webering (1953, pp. 10ff.).

[8] Cf. Crusius (1743, §33) and Crusius (1747, §§524 and 526).

[9] "*Omnis enim a noto ad ignotum scientificus progressus vel a causa est ad effectum, vel ab effectu ad causam; illa quidem est methodus demonstrativa, haec autem resolutiva; alius processus, qui certam rei notitiam pariat, non datur*" (J. Zabarella, *De methodis* III 17, in Zabarella (1597, col. 265F–266A); cf. *De methodis* III 4, ibid., col. 230Ef.; III 18, ibid., col. 266Bff.).

[10] Cf. G. Galilei, *Dialogo sopra i due massimi sistemi del mondo* I, in Galilei (1890–1909, 7, pp. 75f.); letter of June 5, 1637 to Pietro Carcavy, op. cit., 17, pp. 88ff.; *Risposta alle Opposizioni del S. Ludovico delle Colombe e del S. Vincenzio di Grazia*, op. cit., 4, p. 521. For the following development see Mittelstrass (1970, pp. 182ff.).

[11] Letter of November 15, 1638 to M. Mersenne, in Descartes (1897–1910, 2, p. 433).

[12] Cf. Mittelstrass (1970, pp. 294ff.).

[13] Leibniz (1875–90, 7, p. 331).

[14] *Vérités nécessaires et contingentes*, in Leibniz (1903, p. 17).

[15] *Primae veritates*, in Leibniz (1903, p. 518).

[16] *De contingentia*, in Leibniz (1948, p. 304).

[17] Leibniz (1948, p. 301).

[18] Wolff ([3]1740, §663).

[19] *Kritik der reinen Vernunft* B 10.

[20] Ibid. B 2.

[21] Ibid. J. H. Lambert had already made the same distinction: see Lambert (1764, 1, pp. 412ff. [§§634ff.]). Lambert defines as *a priori* a proposition, "wobey wir der Erfahrung vollends nichts zu danken haben" (ibid., p. 414 [§639]), and as *a posteriori* "in sofern wir Erfahrungssätze dazu gebrauchen" (ibid., p. 415 [§642]).

[22] *Kritik der reinen Vernunft* B 11.

[23] Cf. *Metaphysische Anfangsgründe der Naturwissenschaft* A XIII.

[24] Husserl (^{2}1962, p. 28).

[25] Ibid., pp. 20ff.

[26] Cf. Lorenzen (1961), Janich (1973), and Janich (1976).

[27] For a more detailed analysis see Kambartel (1973), Mittelstrass (1974, pp. 56–83, 221–229), and Mittelstrass (1977).

BIBLIOGRAPHY

Crusius, Ch. A.: 1743, *Dissertatio philosophica de usu et limitibus principii rationis determinantis vulgo sufficientis*, Leipzig.

Crusius, Ch. A.: 1747, *Weg zur Gewißheit und Zuverlässigkeit der menschlichen Erkenntnis*, Leipzig.

Descartes, R.: 1897–1910, *Oeuvres de Descartes*, ed. Ch. Adam and P. Tannery, Librairie Philosophique J. Vrin, Paris.

Galilei, G.: 1890–1909, *Le opere di Galileo Galilei, Edizione Nazionale*, Vols. 1–20, Florenz.

Husserl, E.: ^{2}1962, *Die Krisis der europäischen Wissenschaften und die transzendentale Philosophie* (Husserliana 6), ed. W. Biemel, Martinus Nijhoff, Haag.

Janich, P.: 1973, 'Eindeutigkeit, Konsistenz und methodische Ordnung: normative versus deskriptive Wissenschaftstheorie zur Physik' in F. Kambartel and J. Mittelstrass (eds.), *Zum normativen Fundament der Wissenschaft*, Athenäum Verlag, Frankfurt, 1973, pp. 131–158.

Janich, P.: 1976, 'Zur Protophysik des Raumes', in G. Böhme (ed.), *Protophysik*, Suhrkamp Verlag, Frankfurt, 1976, pp. 83–130.

Kambartel, F.: 1973, 'Wie abhängig ist die Physik von Erfahrung und Geschichte? – Zur methodischen Ordnung apriorischer und empirischer Elemente in der Naturwissenschaft', in K. Hübner and A. Menne (eds.), *Natur und Geschichte* (X. Deutscher Kongreß für Philosophie Kiel, 8.–12. Oktober 1972), Felix Meiner Verlag, Hamburg, 1973, pp. 154–169.

Lambert, J. H.: 1764, *Neues Organon oder Gedanken über die Erforschung und Bezeichnung des Wahren und dessen Unterscheidung vom Irrthum und Schein*, Vols. 1–2, Leipzig.

Leibniz, G. W.: 1875–90, *Die philosophischen Schriften von G. W. Leibniz*, ed. C. I. Gerhardt, Berlin.

Leibniz, G. W.: 1903, *Opuscules et fragments inédits de Leibniz*, ed. L. Couturat, Presses Universitaires de France, Paris.

Leibniz, G. W.: 1948, *Textes inédits*, ed. G. Grua, Presses Universitaires de France, Paris.

Lorenzen, P.: 1961, 'Das Begründungsproblem der Geometrie als Wissenschaft der räumlichen Ordnung', *Philosophia Naturalis* 6, 415–431.

Mittelstrass, J.: 1970, *Neuzeit und Aufklärung. Studien zur Entstehung der neuzeitlichen Wissenschaft und Philosophie,* Walter de Gruyter & Co., Berlin and New York.

Mittelstrass, J.: 1974, *Die Möglichkeit von Wissenschaft,* Suhrkamp Verlag, Frankfurt.

Mittelstrass, J.: 1977, 'Historische Analyse und konstruktive Begründung', in K. Lorenz (ed.), *Konstruktionen versus Positionen. Beiträge zur wissenschaftstheoretischen Diskussion,* Walter de Gruyter & Co., Berlin and New York, 1977.

Prantl, C.: 1855–70, *Geschichte der Logik,* Vols. 1–4, Leipzig.

Schepers, H.: 1971, 'A priori/a posteriori', in J. Ritter (ed.), *Historisches Wörterbuch der Philosophie,* Vol. 1, Schwabe & Co. Verlag, Basel and Stuttgart, 1971, col. 462–467.

Webering, D.: 1953, *Theory of Demonstration According to William Ockham* (Franciscan Institute Publications, Phil. Ser. No. 10), St. Bonaventure and Louvain and Paderborn.

Wolff, Ch.: [3]1740 ([1]1728), *Philosophia rationalis sive Logica, methodo scientifica pertractata,* Frankfurt and Leipzig.

Zabarella, J.: 1597, *Opera logica,* Köln.

COMPETING AND COMPLEMENTARY PATTERNS OF EXPLANATION IN SOCIAL SCIENCE*

1. THE STUDY OF EXPLANATORY PATTERNS

At the beginning of his recent book, *The Mathematics of Collective Action*, James Coleman distinguishes two "quite different streams of work in the study of social action" (Coleman, 1973, p. 1). Within the first stream man's behaviour is explained as response to his environment. Causal factors and processes are sought, in that the behaviour is seen as a response called forth by some more or less complicated stimulus. The second stream conceives of behaviour as pursuit of goals, as actions to be explained according to quite a different pattern. Coleman adds in a note that in addition to these two directions of work, which begin at the level of the individual, there are "those, like sociological functionalism and certain parts of ecology, that begin at the level of the collectivity" (Coleman, 1973, p. 1, n. 1).

I don't think he would claim in this way to have established an exhaustive, fourfold classification of approaches in social theory. At least, if we go back a little in time it is fairly easy to identify further approaches, each using a specific pattern of explanation or rather a set of kindred patterns. It has justly been pointed out by Gestalt psychologists[1] that the stimulus-response pattern typical of behaviourism has important features in common with, and may be said to have taken the place, of the associationism which not only dominated the Enlightenment conception of man but also formed a theoretical backbone of experimental psychology in its initial phase. At the level of the collectivity, to use Coleman's expression, one might mention the kinds of explanatory patterns used in the various forms of evolutionism.

To the historian of the philosophy of science the analysis of such patterns may be of value for more than one reason. Some of them obviously point to interesting influences from natural science on the basic conceptions in the study of man and society. That the flourishing of associationism during the Enlightenment was not unconnected with the development of mechanics in the seventeenth century is witnessed

Butts and Hintikka (eds.), Historical and Philosophical Dimensions of Logic, Methodology and Philosophy of Science, 129–158.

by David Hume who explicitly compared association to gravitation. And Hume was not the only one who aspired to be a Newton of the social sciences.[2] On the other hand, in his age and later, much theorizing concerning man and society was still dominated by patterns derived from common sense conceptions. These patterns have as a common core the idea that events rise from the reason and will of human or divine beings. This idea also dominated earlier conceptions of natural processes, those embodied in myths as well as those of philosophical and theological speculations. Ernst Topisch (1958, pp. 3, 5–32) has suggested that the explanatory patterns employed in these forms of thought may be divided into two groups, those derived from human intentional action and those derived from life processes: conception, birth, growth, decay, and death. Among the former he then distinguishes between those based on intentional handling of material things (technomorphy) and those based on social relations and processes (sociomorphy).

I take it that the mechanistic patterns whose success in seventeenth century physics made them so attractive belong to the technomorphic group. They are arrived at by isolating the peripheral part of the handling process: the pressing, pushing or pulling of objects by hand or other parts of the body and the transmission of the effect from the first object (e.g. billiard cue) to further objects (the balls).[3] The central part of the process: the motives, the deliberation, and the decision, which are so important in other technomorphic patterns–those which Topitsch calls intentional–are neglected. Some of these other patterns contain the ideas of divine, creative or regulatory activity that were combined with the mechanistic conceptions in the theistic and deistic cosmologies of the age, God and his work being conceived in analogy with human beings and their intentional products.

The debate about action at a distance provoked by the Newtonian theory of gravitation may be seen as a competition between two different technomorphic patterns, the push-and-pull-pattern being so much more familiar, that many, including Newton himself, balked at the thought of "brute matter ··· (operating) upon and affecting other matter without mutual contact".[4]

This illustrates a point I particularly wish to make: the intimate connection between the explanatory patterns employed in scientific

and proto-scientific theories on the one hand, and the philosophical—
or, if you wish, meta-scientific–conceptions of what constitutes legitimate methods, on the other. For the historian of philosophy of science
not only the explicit pronouncements of epistemological theses and
related requirements of what may pass as an accepted and valid
explanation are of interest. Also the points of view implicit in the
choice of explanatory patterns by natural and social scientists deserve
to be extracted by analysis of the actual conceptions, theories, and
arguments put forward in their work. On the one hand such more or
less explicit epistemological and meta-scientific points of view guide
the scientists in their work. On the other, difficulties in fulfilling some
of the requirements in a fruitful way may lead to their modification or
abandonment. This may take the form of what has been called scientific revolutions, and it is significant that Thomas Kuhn in his account
of one of the main senses of his central term: a *paradigm*[5]–the one he
names disciplinary matrix (in his 'Postscript', 1969)–includes 'values',
i.e. criteria for evaluation of, and choice between, various theories, as
well as what he calls "metaphysical paradigms or models". These latter
turn out to be what I call explanatory patterns. As the values concern
the choice of such patterns, Kuhn seems to go even further than I am
prepared to in connecting the pattern used with the meta-scientific
point-of-view, combining them in a single concept.

2. THE ACTION APPROACH

It would take us too far to inquire into the sources of our various
conceptions of what constitutes an explanation. I take it that these
conceptions all have as a common core the idea that explaining is
connecting, that something is explained if it is shown to fit into a
certain pattern together with other items. In particular, scientific
explanations involve the bringing to light of connections unknown to
common sense. This may show us new ways of changing things
according to our wish–in any case it constitutes an advance in knowledge and understanding and gives a particular kind of satisfaction.

Explanations may be divided according to a number of principles so
as to form a multi-dimensional classification. In a paper read at the

Bucharest-congress I tried to sort out some major dimensions. One has to do with the *strength* of the connection, that is how strongly the explanandum is determined by the explanans. Another concerns the *category* of the item which is explained, e.g. whether it is an event, a state, or a regularity. I also tried to specify a dimension I named the *depth* of the explanation. Some authors have distinguished between explanation by law, by theory and by concept–it is that sort of thing I have in mind as constituting differences in depth.

Here I shall concentrate on a fourth dimension, which has to do with the *kind of pattern* employed. As previously mentioned some of these derive from ordinary, daily life conceptions of what goes on in society, in particular the conception that what goes on is that various individuals do various things, i.e. perform various actions. The term "The action approach" has been used for the approach to social theory based on this conception.[6] It is this approach that is embodied in Coleman's second stream. It must, however, be admitted that his conceptualization, built around the concepts of 'interest', 'control' and 'outcome' (Coleman, 1974, pp. 1–2) is only one among several possible. The common sense conception of purposive action can for scientific purposes be specified in various directions. In its simplest form an individual actor has a given goal that can be specified as a state of affairs to be brought about and his action is explained as the means, it being presupposed that the actor knows or believes that there is such a connection between his action and the goal that the first has a chance of bringing about the latter. It is further presupposed that this is the reason for his action. The simplest explanation of the action is the pointing out of the goal; a more complicated one draws in the actor's relevant beliefs. As Talcott Parsons has stressed, the choice of a means to an end is not always a simple affair solely involving consideration of efficiency. In addition to a norm prescribing the use of the most efficient means, other norms may be involved. If not, Dean Swift's "modest proposal" for solving the population problem by eating roasted babies might be considered justified.

By including all relevant norms and distinguishing between those parts of the environment that the actor is able to influence, and those he is not, we arrive at the pattern that Parsons presented in *The Structure of Social Action* as that of a unit act (Parsons, 1949, pp. 44

and 77).[7] As the title of Parsons' work suggests he was not interested in single acts and their explanation, but in social phenomena and their explanation. His point of departure was that some social theorists had thought it possible to explain social structures and processes as the result of a great number of single acts. These utilitarians, as he called them, were of the opinion that one need not ask what determines the goals in all these acts; a theory explaining the emerging social pattern can be constructed just by presupposing the goals to be distributed randomly (Parsons, 1949, pp. 59–60). Parsons found this – and other aspects of utilitarianism – unacceptable.

Now, those philosophers who really called themselves utilitarians did not think that goals are chosen at random. Instead they were hedonists and associationists, which means that they considered pleasure – and the absence of pain – the general goal of all activity and that they thought that what gives pleasure is a question of what associations have been formed by one's previous experiences. With such a theory a particular action may be explained without its concrete goal being taken as given. Not only the choice of means but also the choice of the goal may be explained – the goal being shown to be what the actor thinks will give him maximum pleasure. Instead of speaking of goals, it is, on this theory, more correct to speak of the outcome, or rather, outcomes of the action. That the actor chose to act as he did may then be explained by showing that he expected it to have outcomes at least as pleasant, everything considered, as any possible alternative. This pattern may also be used by non-hedonists, only it is necessary to presuppose the actor to be able to assign values to outcomes in a consistent way.

What we have now arrived at is a pattern of explanation of individual actions not only well-known in economic theory, but also in other, more general theories of choice and decision. Within the action approach these theories form one important branch of theoretical development. I shall in a moment return to this branch. Before doing so I must mention an important feature of Parson's pattern which points in another direction. As he himself emphasizes, the description of an action and its aspects has to be from the point of view of the actor. It is his beliefs concerning causal connections, not the actual connections as such, that enter into the pattern as one of the determining factors. (Parsons, 1949, p. 46.) We must abandon the naïve realism of some social theorists and acknowledge that the way an individual structures the world he lives in is of paramount importance in understanding and explaining his actions.

In stressing the subjective point of view Parsons aligns himself with Max Weber. Others who also take Weber as a point of departure have gone further in emphasizing the cultural structuring of the 'life world', notably Alfred Schutz and his American pupils.[8] In his critique of the ordinary systems-oriented organization theory David Silverman formulates an action approach of this kind which among its basic tenets are the following: "Action arises out of meanings which define social reality. Meanings are given to men by their society. Shared orientations become institutionalized and are experienced by later generations as social facts ···. Explanations of human actions must take account of the meanings which those concerned assign to their acts; the manner in which the everyday world is socially constructed yet perceived as real and routine becomes a crucial concern of sociological analysis". (Silverman, 1970, p. 127.) As pointed out by Leon Goldstein such a phenomenological approach is deficient as a basis for social theory until it is supplemented by a theory of social change; social theory cannot take the meanings and the structure of the everyday world for granted but must be prepared to explain their origin and development.[9] In fact such a supplementary theory has been sketched in Berger and Luckman, *The Constitution of Social Reality* from 1966, which is in addition to Alfred Schutz's works, the main expression of this approach.

Related to this phenomenological approach is the Gestalt psychological theory of motivation.[10] Both tend to see the situation of the actor as a complex, dynamic field in which he moves according to something like the law of least resistance.

Although this point of view is not excluded from dealing with social relations and social action – as shown in the work of Kurt Lewin – most attempts to develop social theory within the action approach belong to the other main branch, that which includes decision and game theory. Here again the starting point, of course, is the individual actor choosing his course of action according to his preferences. But usually the situation is described in a naïvely realistic and oversimplified way.

Although the term decision theory is of recent origin, this point of view is well-known in economic theory, at least since Adam Smith (Suppes, 1968, pp. 278–79). It is, I take is, not necessary here to present classical decision and game theory.[11] Let me therefore just emphasize some points. These theories have been developed predominantly as normative or prescriptive theories, telling decision makers how to figure out what to do in various situations. As explanatory theories, they accordingly suffer from some deficiencies. First, the presuppositions are rather unrealistic. In the simple forms, at least, the actors are attributed a kind of omniscience and a rationality in their preferences and calculations that at best are approximated in sophisticated organizations. Secondly, in that branch of these theories most relevant to the description and explanation of what is going on between individuals, groups, and organizations in modern society, viz. game theory, viable results for a long time were only obtained in dealing with zero-sum two-party games, whereas most interaction situations involve more than two parties and are of a mixed character, i.e. the interests of the parties are neither completely opposed, nor completely congruent.

There has, it must be admitted, been some progress in these respects in the last couple of decades. It has even been claimed that the limitations can all be overcome.[12] I am not competent to judge this claim, but even if it should turn out that solutions can be found for all classes of games, this whole branch of the action approach seems to me to suffer from a certain one-sidedness. It neglects the role of norms in the determination of actions.

As Parsons stressed in his criticism of utilitarianism, the choice of the means for realizing a given end is not only made in the light of its ability in this respect, but other norms than "the rational norm of efficiency" usually enter the picture (Parsons, 1949, p. 56).

The importance of rules for human action has been emphasized, too, from quite another side. The philosophy of action inspired by Wittgenstein has developed what Peters has called "the purposive rule-following model of behaviour" (Peters, 1958, p. 6). One might illustrate the situation by speaking of two tendencies, the lawyer's tendency to view human actions as determined by norms and the economist's to see them as aimed at realizing values. In sociology the former tendency was formerly dominant but recently "The use of economic points of view" has gained ground "in Sociology", to complete the title of a paper by a Norwegian sociologist from 1967 (Seierstad, 1967, p. 292). It is possible to reduce the basic concepts in each of these approaches to those of the other. This seems advantageous from the point of view of conceptual parsimony, but the price is a high degree of artificiality, and I surmise that a satisfactory theory has to include both sets of concepts.[13]

3. OTHER TYPES OF PATTERNS

The explanatory patterns discussed so far all belong to the action approach. Following Coleman we may oppose them, on the one hand, to those on the individual level explaining behaviour as a more or less automatic reaction to stimuli and, on the other, to patterns on the social level. In the former category we should not only include behaviouristic S–R-patterns, but also those employed in neurophysiological explanations. Finally, I think that associationist patterns belong to this category. Historically they are, from the point of view of social theory, undoubtedly the most important in this category, neurophysiological or S−R-explanations being limited, except in rather loose speculative efforts, to simple kinds of behaviour. We shall take a further look at associationism later.

Functionalism, which Coleman correctly puts in the first place among the social level patterns, may be conceived as an outcrop of what Topitz called biomorphical thought; the organism analogy certainly has played its part in the development of functionalist theory through the ages. I wonder, however, whether it should not, rather, be seen as the heir of that important branch of technomorphical thought explaining things as the result of divine power. The Enlightenment tendency to substitute 'Nature' for 'God' as a principle for explaining order worked then and later in the field of human and social science, in the direction of postulating a 'human nature' which could account for social co-operation and cohesion. Since I intend to return to Enlightenment functionalism in a moment, I shall not pursue the matter further right now, but rather mention some other types of explanatory patterns of importance in social thought and social science.

Topitz characterized, you may remember, biomorphical thought as the use of concepts like conception, birth, growth, decay, and death outside the field of living nature where they belong. Hence the idea that societies or cultures are born, and grow, i.e. develop according to a specific plan, is a typically biomorphical conception. Since evolutionism may take and has taken many different forms there is no single evolutionary pattern of explanation, but rather a whole family.

Although I do not pretend to present an exhaustive classification of explanatory patterns I might conclude my brief survey by mentioning that structuralism should also be considered as having its own kind or family of patterns. From many points-of-view these are extremely interesting and deserve extensive analysis, but since I shall not deal with them in the following I must refrain from even hinting at their specific character.

4. COMPETITION AND COMPLEMENTARITY AMONG PATTERNS

It is quite clear that the various patterns and types of patterns distinguished in the preceding presentation have competed, in the sense that within both prescientific and scientific theory construction the same general phenomena have been dealt with now according to one pattern, now according to another. The various schools or approaches that have fought and criticized each other through the development of Western thinking about man and society have each had their favourite patterns and these have been, or at least seemed to be, incompatible. The quarrels between evolutionists, functionalists, and structuralists have marked the history of anthropology for more than a century; psychologists have been at odds with each other concerning the relative merits of neurophysiological and action-approach-explanations of human behaviour; while philosophers have debated whether these two kinds of explanation exclude each other or not. Quite apart from the contribution an answer to this latter question may render to clarification of the debates of the psychologists, it has a philosophical interest of its own, because it connects with all kinds of fundamental problems within the philosophy of science, notably that of the logic of reduction. This much at least is borne out by the debate between Normal Malcolm and others concerning 'incompatibility' and 'displacement'.[14]

Patterns of explanation may of course relate to each other in other ways than as competitors. They may, for example, complement each other. By this I do not – as one might perhaps expect of a compatriot of Niels Bohr – mean anything very profound. Perhaps supplement would have been a better word – but then I would have missed the alliteration in the title. Now, two patterns may complement or supplement each other in various ways. Let me give an example of one of these ways. A friend of mine has investigated the adoption practice on a small island in the South Seas (Monberg, 1970). Although there are less than eight hundred inhabitants on the island, adoption is so common that he has been able to gather enough material to test hypotheses about the factors determining who adopts whom. It turns out, not unexpectedly, that the best way of explaining the adoption of a child by a couple or a single adult was by referring first to the social norms prohibiting, for example, the adoption of children of one's classificatory brother or sister, and secondly to common preferences. Within the range of choice specified by the norms a boy was preferred to a girl, a child with whose parents one is related to a child of a no-kinsman, and so on. Here we have two patterns, a 'norm-pattern' and a 'value or preference pattern' which co-operate in that the explicandum is seen as partly determined by its position in each, but fully explained when they are combined.

Another kind of complementarity would be represented if we said: All right there are on the island these taboos against adopting certain kinds of relatives, and they, of course, enter the explanation of the actual practice of adoption. But why are these taboos there? It is very probable that what explanation may be discovered follows quite another pattern, for example, a functionalist one. Here we arrive at the concept of reduction. In case we have a theory using one kind of pattern for explaining a range of phenomena and this theory is shown to be reducible to another using a different pattern – take the standard example of phenomenological thermodynamics and statistical mechanics – the latter pattern may also be said to supplement or complement the former.

A third case prevails when a group of phenomena can be explained in two different ways, not only partially as in the adoption case, but fully, but neither theory seems reducible to the other. The phenomena than appear overdetermined and one is tempted to speak of a pre-established harmony.

An example of this constitutes the first of three case studies of complementarity, which make up the rest of my presentation. They are drawn from the eighteenth, nineteenth, and twentieth century, respectively. In the first two explanatory patterns work on two levels, those which Coleman called the individual and the collective level. In social life we may, of course, distinguish more than two levels of organization. On the one extreme we have that of nervous processes, on the other those of a whole society, a civilization or culture, or the whole of mankind. Between these we have the level of the individual person and those of various types of groups and organizations.

Nevertheless, there has been a tendency to consider only two levels, that of the individual, and what is commonly called the social level, but which often is the level of the state. Before embarking on the first case study we have to inquire why this tendency was so prominent in the eighteenth century.

5. THE TWO-LEVEL APPROACH AND THE PROBLEM OF EGOISM

In dealing with social thought we must never forget that through the ages it has been predominantly normatively or prescriptively oriented. The problems whose solution were sought concerned vices and virtues, duties and rights, the principles to be followed in our private and public life, and the values to be pursued. In formulating and in answering such questions various assumptions about the facts of human life and the nature and function of social and political organizations are made. It is the elaborations and criticisms of such assumptions out of which social science has gradually developed[15] as it became recognized that such matters can be dealt with by "the Experimental Method of Reasoning" – to quote the subtitle of the *Treatise of Human Nature*. The idea 'That Politics May be Reduced to a Science' – to use another of Hume's titles[16] – belongs, as I mentioned at the beginning, to the Enlightenment and was prompted by the spectacular advances of natural science since Galileo. Despite the famous criticism in Hume's *Treatise*, Book III, Chapter I, of the "vulgar system of morality" which goes from 'is' to 'ought' neither he nor any other in his time seems quite clear about the relation between fact-finding and explanation on the one hand and normative discourse on the other.[17]

The formulation of the normative questions to which alleged scientific data and principles were supposed to provide answers, was of course affected by historical development – notably by the growth of individualism and the emergence of the national state in the preceding couple of centuries. Accordingly the moral questions about our relations to our fellows tended to be narrowed down to this: Why should one individual care about others, how can the virtue of benevolence and the duty to unselfishness be justified? Analogously, in the field of politics the main question became: Why shall I follow the law of my country, how can the duty of political obedience be justified? Notice how both questions take the individual and his interests as the given point of departure, and how the social level is represented by the state.[18]

This way of putting the questions dominates Hobbes' thought. We all know how he developed answers to both the moral and the political question from a conception of man as inherently egoistical by use of the fiction of a social contract. The theory satisfied few in his time, and today we are aware of its fundamental weakness. To Hobbes' claim that it is rational for each individual to accept the contract and obey the sovereign Parsons rightly objects that this "involves stretching ⋯ the conception of rationality beyond its scope ⋯ to a point where the actors come to realize the situation as a whole instead of pursuing their own ends ⋯" (Parsons, 1949, p. 93).[19] The point is that one can never be sure that it will not be to the advantage of one of the citizens to use 'force or fraud', and if he does that, the whole order may break down. Perhaps the best refutation of this kind of theory is found in Mancur Olson's critique of the idea that it is always rational for the individual member of a group to work for the common interests of the group (Olson, 1971, passim). Even though Olson does not mention Hobbes, it is relevant to the hobbesian theory to point out, as he does, the rationality of the individual who tries to obtain a 'free ride', letting the others share the burden of co-operation and concentrating on his own immediate interest.

In the century following the publication of 'Leviathan' this point was perhaps not so clearly understood; at least the common criticism of Hobbes was rather that he was wrong in taking humans to be so purely selfish, neglecting the benevolent tendencies present in some degree in all of us. Unquestionably the best contribution to the great debate on egoism elicited by Hobbes[20] was bishop Butler's exposition of the structure of the active tendencies in the human mind and analysis of the concepts of 'self-love' and 'benevolence'.[21] A more sensational contribution was Mandeville's *Fable of the Bees* with its thesis of "Private vices, public benefits", meaning that the so-called egoistic pursuance of private interests at the same time benefits the community.[22] Even Butler had not been very far from maintaining that the voice of the enlightened self-love prompts to actions conforming to the dictates of conscience,[23] but Mandeville argued in a different way, pointing out that vices like avarice, prodigality, and pride motivates actions beneficial to society by insuring a sufficient demand for luxuries and other consumer goods to keep the economy growing and by enticing people to public service for the sake of the attached honour.

6. INVISIBLE HAND EXPLANATIONS

Despite the fact that Mandeville was attacked from all sides his main idea that the pursuit of selfish interests may serve the welfare of the community was taken up by some of his attackers. Of special interest here is the group of thinkers commonly labelled 'the Scottish moralists', among whom Francis Hutchison, Adam Smith and Adam Ferguson, in addition to the greatest of them, David Hume, may be mentioned.[24] They agreed in rejecting all attempts to explain – and justify – social and political organization on selfishness or any other single principle. They accepted Butler's analysis of the human mind as involving a multitude of propensities regulated by superior faculties like self-love and conscience.[25] Hume in the *Treatise* attempted to explain unselfish action and moral judgment on an associationist theory of sympathy not far removed from that of Hobbes, but in the second *Enquiry* he had given that up and treated sympathy as a basic ingredient.[26] Adam Smith again developed another concept of sympathy in his remarkably modern psychology in *The Theory of Moral Sentiment*.[27] A third point of agreement among the Scots was their rejection of the social contract theory and its "wild speculations" of "What man was in some imaginary state of nature" (Ferguson, 1966, p. 2). Ferguson quotes Montesquieu – to whom the Scottish Moralists owed much – as saying "Man is born in society, and there he remains" (op. cit., p. 16). Finally, we find in particular in the two Adams, Smith and Ferguson, the Mandevillian idea that each person's pursuit of his own selfish interests may serve the community.

Ferguson not only uses the expression that "nations stumble upon establishments, which are indeed the result of human action, but not the execution of any human design" (Ferguson, 1966, p. 122), but it is a common feature of his description of the origin of social and political institutions that they are very seldom consciously constructed, even though they may serve very important social needs. That they are nevertheless "the result of human action" means that they are unforeseen remote consequences of a multitude of human acts directed at other, more proximate ends, which may be quite selfish.

Adam Smith's famous simile of the invisible hand expresses the same idea. It occurs in both of his main works. In the "Theory of Moral Sentiments" in the form that the rich "are led by an invisible hand" to a nearly equal "distribution of the necessaries of life ⋯ and thus, without intending, without knowing it, advance the interest of society ⋯" (Schneider, 1967, p. 106). In *The Wealth of Nations* it is a question of the market mechanisms leading to an optimal distribution of resources, so that "every individual necessarily labors to render the annual revenue of society as great as he can. He generally indeed neither intends to promote the public interest nor knows how he is promoting it ⋯ he intends only his own gain, and he is in this as in many other cases, led by an invisible hand to promote an end which is no part in his intention" (Schneider, 1967, pp. 106–107). This kind of natural identity of interests, to use Elie Halévy's term,[28] is due, basically, to the simple fact, that in a fair economic exchange between two persons who act rationally both gain by the exchange and each of them, therefore, by taking part in the exchange adds to the total welfare both by gaining himself and by allowing the other to gain.

In Hume's works we not only find analogous formulations, but his whole theory of justice as an artificial virtue is an elaboration of the idea that behaviour may be regulated to the benefit of society by social rules not having any direct source in our active propensities, nor being the result of deliberate planning and agreement.

The idea expressed in the invisible hand simile and in Ferguson's phrase of 'the result of human action but not of human design', I interpret as a clear case of complementarity, a functionalist explanation on the social level – the institutional behaviour in question is explained by its social usefulness[29] – is accompanied by an action approach explanation according to an optimalization pattern: the individual acts so as to maximize the satisfaction of selfish motives. But does not this double explanation of the same phenomena involve a mystical pre-established harmony?

What is in fact the relation between the two explanations? Often the Scots use expressions like "the wisdom of our Maker" (Schneider, 1963, p. 158), or "the wisdom of nature" to hint at a super-natural or an obscure natural cause of the apparently pre-established harmony. But we also find in their work, particularly in that of Hume, suggestions of another account, an account which one and two centuries later were elaborated by the economists Carl Menger and Friedrich v. Hayek, respectively. Menger deals with the question in book three of his methodological treatise from 1883 which was translated into English only a decade ago under the title *Problems of Economics and Sociology* (Menger, 1963). The book is entitled "The Organic Understanding of Social Phenomena", thus alluding to the organism analogy to be found in much functionalism. Here Menger gives a fully spelled out defence of the invisible-hand idea and presents its methodological consequences. Hayek follows suit in two essays published in his *Studies in Philosophy, Politics, and Economics* (Hayek, 1967).[30] Particularly in the first essay, entitled "Notes on the Evolution of Systems of Rules of Conduct" he elaborates the idea which he finds hinted at also by Ferguson and Smith that the harmony is due to a kind of natural selection. The reason why individuals by following certain rules contribute to the preservation and development of society without intending so, is that rules that have this character are favoured by a selection process; societies which develop rules that are not functional in this sense simply succumb in the struggle for survival.

That order may be the result of natural selection among spontaneously emerging variations rather than of design was, of course the main idea in Hume's attack on the argument from design in the *Dialogues on Natural Religion*. Hayek even suggests a historical connection between Hume and Darwin through Erasmus Darwin (Hayek, 1967, p. 119 n. 53).

He summarizes the theory he has found in embryo at Mandeville and the Scots in 9 points. Allow me to quote no. 3 and 4: "It is the resulting overall order of actions but not the regularity of the actions of the separate individuals as such which is important for the preservation of the group." (Hayek, 1967, p. 68.) "The evolutionary selection of different rules of individual conduct operates through the viability of the order it will produce". It is to be observed that the explanatory pattern on the individual level is here the rule-following one, but Hayek is fully aware (see his point 6), that "The concrete individual action will always be the joint effect of internal impulses ··· the particular external events acting upon the individual ··· at the rules applicable to the situation" (id. loc.). It is, in his point of view, less important whether the individual tend to act selfishly or not. Here is, of course, a major difference from Mandeville.

The advantage of invisible hand explanations has recently been stressed by Robert Nozick, who gives a long list of such explanations gathered from old and new social scientific theories and studies. He makes the interesting observation that they seem to fall in two classes: some use the idea of filtering mechanism, some that of an equilibrium process. Filtering here means the same as selection. In many cases Nozick finds, these two are combined, for example, "there might ··· be a filter that eliminates deviations from the pattern that are too great to be brought back by the internal equilibrating mechanisms" (Nozick, 1974, p. 22). This, of course, is also an instance of complementarity.

Before leaving this interesting example of pattern complementarity let me just mention that doubts have been expressed whether selective processes analogous to those active in natural selection in biological evolutionary theory would have time enough to produce the phenomena to be explained (Harsanyi, 1968, p. 307).

7. EVOLUTIONARY PATTERNS OF EXPLANATION

By returning to Adam Ferguson we may enter the path to the second example of competition-complementarity relations in social thought, evolutionism. Ferguson's main work deals with the evolution of man through savagery and barbarism to civilization, the "polished state", as he calls it. Evidence on the "rude states" we not only find in the ancient poets and historians, but also – and better – in descriptions of contemporary people who still have not left these stages; "It is in their present condition, that we are to behold, as in a mirror, the features of our own progenitors" (Ferguson, 1966, p. 80). These words contain the programme of the comparative method so basic to all social and cultural evolutionism.[31] In Ferguson we also find the idea of a fixed chronological order of stages through which a nation develops. An alternative conception of development sees this as a more continuous process during which some characteristics become more and their oppositions less prevalent, for example, the Durkheimian development from mechanical to organic solidarity.

Common to all evolutionists is the idea that development is orderly, follows a pattern of stages or a pattern of continuous change in a specific direction.[32] This means that the state at a certain time of a unit of the kind that follows the development pattern in question may be explained by its being at a certain point in the development. Whether or not one wants to talk about such explanations as resting on a particular kind of law, they differ from other explanations in a way which justifies setting up a special category of evolutionary patterns of explanation, as I did in my survey of such patterns. But evolutionists differ, for instance on the kind of evidence by which they support the claim that units of a certain kind always develop in a certain manner. One possibility is to present this as simply a matter of fact established by experience and induction. The law in question, if one uses this term, is then what is often called an empirical generalization. Explanations in terms of such laws are somewhat suspect – at least they are shallow explanations – and there is always a tendency to support them by explaining why the law in question holds. This second explanation usually follows another pattern, often on a lower level of organization. For instance the Scottish moralists tend to refer to human nature as something basic and immutable, containing the seeds of development in that the springs of action of the individuals lead them to a conduct favouring social change in a certain direction. The 19th century blamed them – and the Enlightenment generally – for this belief in an immutable human nature, maintaining instead that evolutionary change is so pervasive that no general feature is found and no law is valid in all stages. How then explain the pattern of development? Idealism resorted to the conception of development as a logical process. Maybe we here have a wholly new kind of explanatory pattern; I really don't know. As a matter of fact, I have only recently begun working in the enormous field of evolutionary social theory. I shall therefore restrict this part of my address to a presentation of the questions my approach leads me to ask concerning evolutionism.

First and foremost it is necessary to delimit the class of evolutionary social theories and group them according to the explanatory pattern employed. This is no simple task. Not every theory of social change is evolutionary, and not every law in which time enters is an evolutionary law.[33] Distinctions like those between special and general evolution[34] and between unilinear and multilinear evolution[35] have been introduced, and the relation between evolutionism and historicism[36] has to be cleared up. Finally there are major differences between the older kind of evolutionary theory which Ferguson represents – the so-called "Conjectural history"[37] on the one hand, and the post-Darwinian theories on the other.

When these matters have been cleared up to some degree the main question concerning the various evolutionary theories can be raised, namely: do they take evolution as a basic fact, or do they contain answers to the question why evolution takes place, and in the latter case, what kinds of explanations do they offer, and what is then the relation between the evolutionary pattern as such and the patterns used in these explanations?

Such questions are raised concerning Marx by Maurice Mandelbaum in his paper on 'Societal laws'. Mandelbaum distinguishes between "functional" and "directional laws" or "laws concerning history" and "laws of history", and says that "Marx' theory of the relation between the economic organisation of a society and other institutions in that society (his doctrine of "the superstructure") would be a functional law, i.e. a law concerning history. On the other hand, one may interpret his view of dialectical development as an attempt to formulate a law stating a necessary pattern of directional change in history, i.e. as a "law of history". In a note he then adds "It appears to me that Marx failed to see the difference between these two types of explanation. To be sure, there are passages in Marx which would make it appear that the pattern of directional change is merely the result of the forces which he attempts to analyse in terms of his economic theory and his doctrine of 'the superstructure'. Nevertheless, it does seem equally plausible (at the very least) to maintain that he regarded the pattern which he traced in the history of mankind as having an inherent necessity in it, i.e. that it was 'irresistible', and was not merely the actual outcome of forces operating from moment to moment" (Mandelbaum, 1957, p. 215).[38]

Although John Stuart Mill is no typical evolutionist he still follows Comte in distinguishing between social statics and social dynamics, and like him he tends to neglect the former. One might even say that he was influenced by Comte's insistence that human nature is historically conditioned and that hence there can be no direct deductions from the laws of human nature, as his father and others had thought. On the other hand he did not doubt that these laws, which he of course conceived in an associationist sense, underlay all social life and evolution. His main motivation in writing the *Logic* had to do with his wish to clear up these matters – to balance his father's deductive method against the historians' inductive method and the Comtean and Saint-Simonean evolutionary conceptions of history. The solution he found was the historical or inverse deductive method: we must first find 'the laws of progress' empirically. Only then can we try to explain them by showing how, after all, they follow from the basic laws of human nature in a complicated way. The fault of James Mill was to believe that we can start from these laws and use the direct deductive method. "The Franch school", on the other hand, made the error of "supposing that the order of succession which we may be able to trace among the different states of society and civilization which history presents to us ··· could ··· amount to a law of nature. It can only be an empirical law. The succession of states of ··· human society cannot have an independent law of its own, it must depend on the psychological and ethological laws which govern the action of circumstances on men and of men on circumstances" (Mill, 1876, vol. 2, p. 512). Not all 19th century social theorists would agree! Many, perhaps most, were historicists, who maintained that the only laws of universal validity for human society are developmental laws, so that these are irreducible. In some we may perhaps find the idea that developmental laws and other laws support each other, not in the way that one kind is reducible to the other, but in a more complex and mutual way. That would be a fourth kind of complementarity of explanatory patterns of great interest.

8. EXCHANGE THEORY

My third, and final, example of complementarity does not, like the other two, involve patterns on different levels – at least not directly. Instead it concerns the relation between the two kinds of action approach patterns mentioned before, namely the kind featuring rules or norms as major ingredients and the optimizing one. There is one particular norm which has been extensively used as an explanatory principle, the norm of reciprocity. We find it mentioned as such in the works of sociologists like Hobhouse and Simmel, and in 1959 Alvin Gouldner read a paper entitled 'The Norm of Reciprocity: A Preliminary Statement' (Gouldner, 1960) at the annual meeting of the American Sociological Association. He thereby started a major theoretical development in contemporary sociology. Different authors interpret the norm somewhat differently, the main idea, however, being that in a situation where one party gives something of value or disvalue to another party this other party should reciprocate by returning something of roughly equal value or disvalue. There does not seem to be full agreement among those who around this concept of reciprocity in exchange have developed what is now called exchange theory, as to how wide an area in social life this theory covers, but at least some tend to see all social interaction as exchange.

Not only sociologists, but also anthropologists, particularly in France,[39] have been interested in exchange and reciprocity, as witnessed by Marcel Mauss' classical *Essay sur le don* from 1925, published in English as *The Gift* (Mauss, 1954). Also in Lévi-Strauss' first major work "Les structures élémentaires de la parenté" (Lévi-Strauss, 1949) he treats marriage rules in terms of exchange of women between lineages and even has a chapter entitled 'Le principe de reciprocité'.[40] It is worth mentioning that one of the main representatives of sociological exchange theory, G. C. Homans, began his preoccupation with exchange by criticizing Lévi-Strauss' theory. I shall, however, not here deal with anthropological exchange theory which seems to differ from the sociological one by taking social groups – lineages etc. – as parties involved, instead of individuals.[41]

In Gouldner's presentation of the norm of reciprocity the point is stressed that the goods exchanged may be of very many different kinds, material and immaterial. Homans has for example studied situations in office work where help by an older and more experienced colleague is repaid by deference and expression of gratitude. We see here a problem: How do we measure the value of these things? In economic exchange we may take the market price as a rough measure of value, but help, deference etc. has no market price in money. Gouldner is aware of the problem, which he calls "the problem of equivalence" (Gouldner, 1960, p. 171). He states that "equivalence ⋯ refers to a definition of the exchangeables made by the actors in the situation" (op. cit., p. 172), which "differs ⋯ from holding that the things exchanged ⋯ will be objectively equal in value ⋯" (id. loc.). But there are two actors, and they may differ in their evaluation! This creates difficulties for the testability of Gouldner's thesis that the norm of reciprocity is universally acknowledged.

Of particular interest in our context is Gouldner's discussion of the relation of egoism to reciprocity. He states that "the reciprocal processes ⋯ actually mobilize egoistic motivations. Benthamite utilitarianism has long understood that egoism can motivate one party to satisfy the expectations of the other, since by doing so he induces the latter to reciprocate and to satisfy his own" (op. cit., p. 173). Gouldner distinguishes between "the existential belief in reciprocity", which "says something like ⋯ 'People will usually help those who help them' ", and the norm, "that people should help those who help them". Both, he says "enlist egoistic motivations in the service of social system stability". On the other hand, when asking why the existential belief tends to be correct ("why men reciprocate gratifications") he answers by referring to the norm. And this norm is an action motivating factor because it is internalized.

We have here first an explanation why a person tends to repay a service or a gift, namely that the person in question follows a norm of reciprocity. This is the rule-following pattern. But Gouldner further presumes the norm-following to be the result of a process of internalization. What pattern of explanation is at work here? Well, there is more than one theory of internalization, and it would take us too far to try to analyse, for example, the Freudian theory here. Suffice it to mention three points. First, the concept of internalization seems to presuppose some kind of inertia – a very important explanatory concept. A structure created by a certain process released by outside influence is conceived as existing continuously, resisting change and channalizing action. Secondly, it is difficult to overestimate the importance of internalization and identification, with which it is combined, at least by Freud, for the understanding of human conduct. One of the reasons why a pure decision-theoretical approach based on self-love is insufficient, is that the self whose gratification is at stake itself is a result of social interaction involving identification and internalization. Finally, since the internalization of a norm presuppose that the norm is there as a part of the culture of the group or the society, exchange theory in Gouldner's presentation may, after all, be said to span two levels.

We could arrive at this point also by taking up Gouldner's remarks about egoism and reciprocity, quoted above. They can be interpreted as signifying that often the action which follows the reciprocity norm is also motivated by expediency, i.e. can also be explained according to a satisfaction-maximation pattern. This would be an instance of the kind of complementarity where two explanations with different patterns point out factors which support and supplement each other.

Gouldner may, however, entertain the further idea that the norm has its origin in the fact that so often the expedient thing to do is to reciprocate. Even if this may not itself give rise to such a social norm, the norm may emerge thanks to the further fact, also stressed by Gouldner, of its stabilizing role in society. We then have an instance of a Hayek-type invisible-hand explanation. And this, of course, is a two-level affair.

Whereas Gouldner in his paper only sketched the exchange theoretical point of view, Homans tried in his book-length treatment (Homans, 1961) and other places (Homans, 1962) to formulate a real theory in accordance with his deductionistic and reductionistic philosophy of social science (Homans, 1967). Here the place of the norm of reciprocity is taken by a 'rule of distributive justice'. This name is misleading – the rule concerns retributive rather than distributive justice – and it is a complex rule; it states that people expect that in an exchange relation "the rewards of each man be proportional to his costs" and that "the net rewards, or profits, of each man be proportional to his investments" (Homans, 1961, p. 75).

The concept of investment covers various background factors that are believed to entitle persons to higher rewards: age, seniority, talent, skill, education, sex, race, etc. These vary from society to society and from group to group (Homans, 1961, p. 246). In particular, some societies stress ascribed factors, others achieved.

Homans gives a number of examples of the working of the rule of distributive justice, showing that it explains quite a range of phenomena. He also suggests that the existence of the rule, that is the occurrence of the expectations involved, may be explained by behaviour in accordance with the rule being in general rewarding. It is not quite clear how such an explanation should be carried through, but Homans is committed to the position that all exchange phenomena in what he calls elementary behaviour can ultimately be explained in terms of operant conditioning.

The theory of such conditioning, mainly developed by Homans' friend B. F. Skinner, uses a rather simple explanatory pattern belonging to the class of S−R-patterns.[42] Homans' position, then, seems to be that explanation of social behaviour in terms of the rule of justice may and should be supported by the existence of this rule being explained in terms of operant conditioning. But he does not show explicitly how this can be done, and serious difficulties in this position have been pointed out by critics. Morton Deutsch, for instance, has analysed the way he uses Skinnerian theory and arrived at the conclusion that "the link between Homans' theory and behaviouristic psychology is brittle" (Deutsch, 1964, p. 163). He also points out that Homans faces the same problem of the measure of value that marred

Gouldner's presentation (op. cit., p. 161). Concerning the rule of justice Peter Blau has questioned its status: "Homans does not explicitly state that the rule is a social norm ⋯ Indeed, he sometimes implies that it is a natural sentiment ⋯" (Blau, 1964b, p. 196). The question of its explanation of course turns out quite differently according to which of these alternatives one accepts. Blau himself thinks that the rule is a social norm. Expectations of fairness in exchange seem in fact to work as a corrective to what 'economic' motivations alone would lead people to do.

A person may abstain from using the power that the scarcity of the goods he commands gives him to drive an unfair bargain, because he accepts the rule of justice. But then such expectations cannot be explained as a direct result of people having such motivations. And these seems to be the ones that follow operant conditioning theory.

Blau himself prefers to acknowledge that power relations enter into social exchange. His exchange theory is accordingly also a theory of social power (Blau, 1964a).

9. CONCLUSION

Of the examples of complementarity I have here presented the two last have this in common that the problem is whether some kind of explanation can and should be reduced to another. Mill, for example, denied that evolutionist explanations can be final and ultimate and demanded that they should be supported by associationist explanations. Hence the explanatory patterns are on two levels. In the case of exchange theory a similar question is raised concerning explanations in terms of a norm which is interpreted as expectations in individuals and explanations of their behaviour in general according to an action approach pattern (Gouldner) or the operant conditioning pattern (Homans) which also work on the individual level.

Invisible hand explanations are more complicated. Here the same phenomena are explained according to patterns on two levels, and a third pattern – Providence or natural selection (filtering) – is introduced to explain the harmony between the levels.

Other case studies of social theory formation should be made to bring other kinds of combination of explanations to light. Such studies may, I believe, prove of value, not only in historical research, but also as a means of furthering integration in social science by elucidating the complicated relations between theories and disciplines as they exist today.

University of Copenhagen

NOTES

* This research has been supported by grants 515–1816 & 515–3629 from the Danish State Research Council for the Humanities.
1 See, e.g. Köhler (1929).
2 Hume's methodological points-of-view and their relation to his understanding of Newton has recently been fully and convincingly dealt with by James Noxon (1973, passim).
3 Topitsch only at the end of his book (1958, pp. 264–66) mentions that the technomorphic models have a "mechanistic component" which was developed in classical atomism, and that modern science in its initial stages was influenced strongly by common-sense notions.
4 Cf. Newton's letter of Bentley (1692–93) quoted by Buchdahl (1969, p. 382).
5 Prompted by his critics, among whom Margaret Masterman (1970, passim) had distinguished 21 different senses in which Kuhn had employed the term 'paradigm', he used the first part of his 'Postscript 1969' to the second edition of his book (Kuhn 1970) to explain the development of the concept in his thinking and to admit that there had been ambiguities in his use of the term. In this context he introduced the new term "disciplinary matrix" to cover one of the original senses of 'paradigm'.
6 The expression 'action approach' has been used by Percy S. Cohen (1968, p. 69) while John Rex writes of "the action frame of reference" (Rex, 1961, p. 78). David Silverman (1970, p. 126) uses both expressions. Although these three authors mean somewhat different things by these expressions they agree in opposing the action approach (frame of reference) to a holistic, functionalistic approach.
   'The Action Approach' has also been used as a translation of Alain Touraine's "action-nalisme" (Touraine, 1964, p. 79). Since a main point in his theory is that the subject of social action is not an empirical subject, in particular not an individual person, we have here a quite different sense.
7 In *Towards a General Theory of Action* Parsons modified his scheme. In addition to the actor an action is there said to have the following elements: ends or goals, situation and orientations. An action is normatively regulated and motivated (Parsons, 1952, pp. 42–43).
8 Schutz (originally Schütz) came to the USA in 1939 as a refugee from Vienna. He taught at the New School for Social Research in New York until he died in 1959. Although not a pupil of Husserl, Schutz since the publication in 1932 of his *Der sinnhafte Aufbau der sozialen Welt* maintained close contact with Husserl who had greeted the work warmly. Among American sociologists who have been influenced by Schutz are Harold Garfinkel, the 'founder' of ethnomethodology, Maurice Natanson and Peter L. Berger.

[9] Cf. Goldstein (1963), p. 296.

[10] Concerning the Gestalt theory of motivation cf. Heider (1960).

[11] The basic presentation was given by von Neumann and Morgenstern in 1944. A more modern, comprehensive text-book is Luce and Raiffa (1957). An important distinction in decision theory is that between decision making under certainty, decision making under risk, and decision making under uncertainty. In the first the outcome of each alternative action is known with certainty, in the second it is only known with a known probability, and in the third the outcome depends on the action of one or more other actors.

[12] Cf. Harsanyi 1968, p. 309. Important advances in game theory have been made by Schelling (1960, passim) and others who have treated 'bargaining-situations'.

[13] Robert Nozick discussed the whole question of rules versus values in quite another context in terms of a difference between a 'side constraints view' and an 'end-state-maximation view' (Nozick, 1974, pp. 28–30). He argues for the importance of the former, although he – in a note (op. cit., p. 29) – mentions the formal possibility of side constraints (norms) being included in the end state to be achieved by treating their transgression as a factor with infinite negative weight. This, of course, is the way an economist wants to come to terms with social norms. Another possibility of reduction is mentioned by Harsanyi, namely to show that the norm in question has originated in people's personal objectives and interests (Harsanyi, 1968, p. 314).

[14] Cf. Malcolm (1967) and (1968), and Bernstein (1972, pp. 281–99).

[15] A normative or prescriptive problem may lead to fact-finding and construction of explanatory theory in many ways. Not only because prescription often presupposes a knowledge about the probable outcome of alternative actions, but also e.g. because a fundamental norm like "act according to nature!" requires investigation of what nature is and how it works.

[16] Cf. Hume (1898, vol. 3, pp. 98–109).

[17] A. MacIntyre has challenged the accepted interpretation of the passage in question according to which it shows a clear understanding of what later came to be called the naturalistic fallacy. MacIntyre's paper, originally published 1959 in Philosophical Review, is reprinted together with some criticisms in Hudson (1969, pp. 38–80).

[18] The tendency in question is also prevalent in economics. It is natural there to take the national state as the large unit, since its government influences economic life directly, and since its boundaries are usually custom lines. On the other hand, the family is often taken as the smaller unit by economists, it being an economic unity.

[19] Cf. also Parsons (1968), passim.

[20] A good, short survey of this debate is given in Hall (1956, pp. 371–82).

[21] Cf. Broad (1930, pp. 60–83), and Duncan-Jones (1952).

[22] Mandeville first (1705) published his poem under the title 'The Grumbling Hive: or, Knaves Turn'd Honest'. In 1714 a second edition enlarged by An Enquiry into the Origin of Moral Virtue was issued under the title The Fable of the Bees, or Private Vices, Public Benefits. New and further enlarged editions were published 1723 and 1724. Cf. Mandeville (1970).

[23] Cf. Broad (1930, p. 80).

[24] Cf. Bryson (1968). Bryson includes in addition to the four mentioned in the text also Thomas Reid, Dugald Stewart, Lord Kames and Lord Monboddo.

[25] See in particular Hume's 'Of self-love' (Appendix to the Second Enquiry) (Hume, 1898, vol. 4, pp. 266–72).

[26] Cf. Noxon (1973, pp. 189–90). Noxon takes this change as a symptom of Hume's

having given up the original plan of creating a science of human nature on associationism.

[27] Cf. Campbell (1971), in particular ch. 4. On the importance of the concept of sympathy see note 28.

[28] By 'the thesis of the natural identify of interests' Halévy understands the principle that "the various egoisms harmonize of their own accord and automatically bring about the good of the specie (Halévy, 1928, p. 15). He mentions two other ways in which egoism was made compatible with a utilitarian view. The 'principle of the fusion of interests' which point to sympathy as a mechanism by which the well-being of others becomes a goal for my egoism. The 'thesis of the artificial identification of interests' is the Benthamist principle that the rulers ought, for instance by the penal code, make public spirited conduct preferable from a selfish point of view (Halévy, 1928, pp. 13–18).

[29] Dugald Stewart, who a generation later than Ferguson and Smith summarized and systematized the main tenets of the Scottish school defends teleological explanations against Bacon's attack. In *Elements of the Philosophy of the Human Mind* he states that the danger of applying such explanations arises mainly in the philosophy of mind, because one there tends to confuse final causes with efficient ones, i.e. to believe that the "happiness and improvement both of the individual and of society", which is "the end for which we may presume" that our mental constitution was "destined by our Maker" is the efficient cause of our conduct. Adam Smith, according to Stewart, never committed this error (Schneider, 1967, pp. 148–58).

[30] The second essay is entitled "The Results of Human Action but not of Human Design" (Hayek, 1967, pp. 96–105). It deals mostly with law as such a result.

[31] For a description and criticism of this method see Nisbet pp. 189–208.

[32] In the article on the concept of evolution in "International Encyclopedia of Social Sciences" R. C. Lewontin sets up "a hierarchy of principles in the evolutionary world vices: change, order, direction, progress, and perfectibility" (Lewontin, 1968, p. 203). I would only include theories which have all of the three first principles.

[33] Cf. Mandelbaum (1971, p. 116).

[34] Cf. Sahlins & Service (eds.) (1960, chap. 2).

[35] Cf. Stewart (1955), passim.

[36] I take 'historicism' in the Popperian sense (Popper, 1961, passim), and not as a translation of German 'Historismus'. Cf. Donegan's illuminating clarification in Donegan (1974). Cf. also Mandelbaum (1971).

[37] Cf. Nisbet (1969, p. 143). Another term used for this kind of philosophy of history is "reasoned history (histoire raisonné)". Cf. Burrow (1966, chap. 3): 'The Reasoned History of Man'.

[38] In his treatment of Marx's 'historicism' in a later work (Mandelbaum, 1971) Mandelbaum arrives at the conclusion that he did not consider directional laws as irreducible (op. cit., p. 72).

[39] Malinowski, however, has played a major role in the development of anthropological exchange theory, mainly by his study of the Kula exchange. And Frazer may be said to have offered the first theory of social exchange. Cf. Ekeh (1974, pp. 20–33).

[40] This chapter is reprinted in English in Coser and Rosenbaum (1964, pp. 74–84).

[41] This is a main point in Ekeh's recent analysis of the two traditions, of which he obviously prefers the collectivist one (Ekeh, 1974, passim).

156   M. BLEGVAD

[42] It is not my intention to deny that there is a difference between Pavlovian and operant conditioning, but they clearly belong to the same approach, different from the action approach.

BIBLIOGRAPHY

Bernstein, R. J.: 1972, *Praxis and Action*, Duckworth, London.
Blau, P. L.: 1964a, *Exchange and Power in Social Life*, John Wiley and Sons, New York.
Blau, P. L.: 1964b, 'Justice in Social Exchange', *Sociological Inquiry* **34**, 193–206.
Broad, C. D.: 1930, *Five Types of Ethical Theory*, Kegan Paul, Trench, Trubner & Co., London.
Bryson, G.: 1968, *Man and Society: The Scottish Inquiry of the Eighteenth Century* (1945), Repr. Augustus M. Kelly, New York.
Buchdahl, G.: 1969, *Metaphysics and the Philosophy of Science*, Blackwell, Oxford.
Burrow, J. W.: 1966, *Evolution and Society. A Study in Victorian Social Theory*, Cambridge Univ. Press.
Campbell, T. D.: 1971, *Adam Smith's Science of Morals*, George Allen & Unwin, London (Univ. of Glasgow Social and Economic Studies, Vol. 21).
Cohen, P. C.: 1968, *Modern Social Theory*, Heinemann, London.
Coleman, J.: 1973, *The Mathematics of Collective Action*, Heineman Educational Books, London.
Coleman, J.: 1974, *Social Structure and a Theory of Action*, (Unpublished paper read at the 8th World Congress of Sociology, Toronto).
Coser, L. A. and Rosenberg, B. (eds.): 1954, *Sociological Theory. A Book of Readings*, 2nd ed., Macmillan Co., New York.
Deutsch, M.: 1964, 'Homans in the Skinner Box', *Sociological Inquiry* **34**, 156–165.
Donegan, A.: 1974, 'Popper's Examination of Historicism', in P. A. Schilpp (ed.): *The Philosophy of Karl Popper*, Open Court, La Salle, Illinois, Vol. 2, pp. 905–24.
Duncan-Jones, A.: 1952, *Butler's Moral Philosophy*, Pelican, Harmondsworth.
Ekeh, P. P.: 1974, *Social Exchange Theory. The Two Traditions*, Heinemann, London.
Ferguson, A.: 1966, *An Essay on the History of Civil Society*, 1767 (ed. by Duncan Forbes), Univ. of Edinburgh Press.
Goldstein, L. J.: 1963, 'The Phenomenological and Naturalistic Approaches to the Social Sciences' (*Methodos* 13, 1961), repr. in M. Natanson (ed.): *Philosophy of the Social Sciences*, Random House, New York, pp. 288–301.
Gouldner, A. W.: 1960, 'The Norm of Reciprocity: A Preliminary Statement', *American Sociological Review* **25**, 161–178.
Halévy, E.: 1952, *The Growth of Philosophical Radicalism* (1901–04), Transl. by Mary Morris (1928). Repr. with corrections, Faber & Faber, London.
Hall, E. W.: 1956, *Modern Science and Human Values*, D. von Nostrand Co., Princeton, N.J.
Harsanyi, J. C.: 1968, 'Individualistic and Functionalistic Explanations in the Light of Game Theory: The Example of Social Status', in I. Lakatos and Musgrave, A. (eds.): *Problems in the Philosophy of Science*, North-Holland Publ. Co., Amsterdam, 1968, pp. 305–21 and 337–48.
Heider, F.: 1960, 'The Gestalt Theory of Motivation', *Nebraska Symposium on Motivation* 1960, pp. 145–172.

Homans, G. C.: 1961, *Social Behaviour, Its Elementary Forms*, Routledge & Kegan Paul, London.

Homans, G. C.: 1962, *Sentiments and Activities. Essays in Social Science.* Routledge & Kegan Paul, London.

Homans, G. C.: 1967, *The Nature of Social Science*, Harcourt, Brace & World, Inc., New York.

Hudson, W. D. (ed.): 1969, *The Is-Ought Question.* Macmillan, London. (Controversies in Philosophy).

Hume, D.: 1898, *The Philosophical Works of* (ed. by T. H. Green and T. H. Grose), Vols. 3–4, New impr. Longmans, Green, and Co., London.

Kuhn, T. S.: 1970, *The Structure of Scientific Revolutions*, 2nd ed. Univ. of Chicago Press, Chicago.

Köhler, W.: 1929, *Gestalt Psychology*, Liveright Publ. Co., New York.

Lewontin, R. C.: 1968, 'The Concept of Evolution', in D. L. Sills (ed.): *International Encyclopedia of the Social Sciences.* Macmillan Co. & The Free Press, Vol. 5, pp. 202–210.

Luce, R. C. and Raiffa, H.: 1957, *Games and Decisions, Introduction and Critical Survey.* John Wiley & Sons, New York.

Malcolm, N.: 1967, 'Explaining Behavior', *Philosophical Review* **76**, 97–104.

Malcolm, N.: 1968, 'Conceivability of Mechanism', *Philosophical Review* **77**, 45–72.

Mandelbaum, M.: 1957, 'Societal Laws', *The British Journal for the Philosophy of Science* **8**, 211–24.

Mandelbaum, M.: 1971, *History, Man & Reason. A Study in Nineteenth-Century Thought*, The Johns Hopkins Press, Baltimore.

Mandeville, B.: 1970, *The Fable of the Bees* (ed. by Phillip Harth), Penguin Books.

Masterman, M.: 1970, 'The Nature of a Paradigm', in I. Lakatos and A. Musgrave (eds.): *Criticism and the Growth of Knowledge*, Cambridge Univ. Press, pp. 59–89.

Mill, J. S.: 1875, *A System of Logic, Ratiocinative and Inductive.* 9th ed., Vols. 1–2. Longmans, Green, Reader, and Dyer, London.

Monberg, T.: 1970, 'Determinants of Choice in Adoption and Fosterage on Bellona Island', *Ethnology* **9**, 99–136.

Nisbet, R. A.: 1969, *Social Change and History. Aspects of the Western Theory of Development*, Oxford Univ. Press, London.

Noxon, J.: 1973, *Hume's Philosophical Development. A Study of his Methods*, The Clarendon Press, Oxford.

Nozick, R.: 1974, *Anarchy, State, and Utopia*, Blackwell, Oxford.

Olson, U.: 1971, *The Logic of Collective Action*, Paperback ed., Harvard Univ. Press, Cambr. Mass.

Parsons, T.: 1949, *The Structure of Social Action*, 2nd ed., The Free Press, Glencoe, Ill.

Parsons, T. and Shils, E. (eds.): 1952, *Toward a General Theory of Action*, Harvard Univ. Press, Cambr. Mass.

Parsons, T.: 1968, 'Order as a Sociological Problem', in P. G. Kuntz (ed.): *The Concept of Order*, Univ. of Washington Press, pp. 373–84.

Peters, R. S.: 1958, *The Concept of Motivation*, Routledge & Kegan Paul, London.

Popper, K. R.: 1961, *The Poverty of Historicism*, Paperback ed. Routledge & Kegan Paul, London.

Rex, J.: 1961, *Key Problems of Sociological Theory*, Routledge & Kegan Paul, London.

Sahlins, M. D. and Service, F. R. (eds.): 1960, *Evolution and Culture*, The Univ. of Michigan Press, Ann Arbor.

Schelling, T. C.: 1963, *The Strategy of Conflict*, Oxford Univ. Press, London.

Schneider, L. (ed.): 1967, *The Scottish Moralists on Human Nature and Society*. Phoenix Books, Univ. of Chicago Press.

Seierstad, S.: 1967, 'Om bruken av økonomiske synspunkter i sosiologien', *Tidsskrift for samfunnsforskning* **8**, 292–304.

Silverman, D.: 1970, *The Theory of Organisations, A Sociological Framework*, Heinemann, London.

Steward, J. H.: 1955, *Theory of Cultural Change. The Methodology of Multilinear Evolution*, Univ. of Illinois Press, Urbana.

Suppes, P.: 1968, 'Information Processing and Choice Behavior', in I. Lakatos and A. Musgrave (eds.): *Problems in the Philosophy of Science*, North-Holland Publ. Co., Amsterdam, 1968, pp. 278–99.

Topitsch, E.: 1958, *Vom Ursprung und Ende der Metaphysik. Eine Studie zum Weltanschauungskritik*, Springer-Verlag, Wien.

Touraine, A.: 1974, 'Towards a Sociology of Action' in A. Giddens (ed.): *Positivism and Sociology*. Heinemann, London, pp. 75–100.

EVANDRO AGAZZI

SUBJECTIVITY, OBJECTIVITY AND ONTOLOGICAL COMMITMENT IN THE EMPIRICAL SCIENCES

It is an easily recognizable feature of the way science is considered in our days that almost no educated person or scientist would claim that science is a *true* knowledge (an exception is sometimes made for mathematical disciplines, but their 'truth' is in any case considered as a very special one and will therefore not be of our concern in the present paper). Such a general tendency not to mix truth with science does not express on the other hand any mistrust or underestimation of the science of our time; on the contrary, it has coincided with a growth of such an estimation up to the point that an equally general tendency is nowadays established to consider science as the model, as the 'paradigm' of our knowledge, in the sense that every field of research in which 'rigourous' knowledge is reached by respecting some accepted methodological requirements, is accepted to constitute a science.

On the other hand, while 'truth' has been recognized as a too engaging requirement to be attributed to science, the simple 'rigour' may have seemed perhaps too modest to qualify it, so that a rather special characteristic is being more and more frequently advocated for it, and this is *objectivity*, which appears to occupy a kind of intermediate position between truth and rigour, inasmuch as it shares all the characters of rigour, without sharing all the characteristics of truth.

But if one tried to find out a commonly admitted characterization of objectivity, one would not succeed very easily in that enterprise for there is rather a good deal of expressed or unexpressed shades of meaning in this concept, which can go from the pure fact of identifying it with the qualification of science as a 'public' discourse, up to considering some structural properties of the relevant scientific statements (like 'invariance' with respect to different frames of reference), up to the qualification of objectivity as meaning more or less directly 'verisimilitude', as Popper says, in the sense of a progressive 'approximation to truth'.

We are not going to discuss these different viewpoints, which are

Butts and Hintikka (eds.), Historical and Philosophical Dimensions of Logic, Methodology and Philosophy of Science,
159–171.
Copyright © 1977 by D. Reidel Publishing Company, Dordrecht-Holland. All Rights Reserved.

frequently connected with some underlying general philosophical perspectives (such as idealism, realism, phenomenalism, etc.), but we shall
rather try to find out whether there are some common features of the
concept of objectivity, which may be identified independently of such
or similar presuppositions.

The first remark that can be used as a starting point of our analysis is
the fact that, already in our everyday language, objectivity is most
frequently meant to express some 'independence of the subject' (in
such a sense, e.g. we qualify as 'objective' especially judgments, ap-
preciations, records of facts, expertises, etc.). Such a fact happens so
frequently and so naturally, that we may not pay any attention to
something which should deserve at least some reflection: viz. that in so
doing we are not keeping faith with what the linguistic root of the term
'objective' ought to imply. In fact, this root would lead us to define
some property, some feature, as objective when it is thought to be
'pertinent to the object' and hence to qualify as objective also statements that describe such a state of affairs. On the contrary, no
reference to that object is made in the most common acceptation of
objectivity, but rather an indirect reference to the 'subject' (in the
sense of an 'independence of' it).

There is no novelty proper in what we have just pointed out and one
could recall here that the common sense use of the notion of objectivity is reminiscent of an epistemological distinction that can be traced
between a 'strong' objectivity (which means 'inherence to the object')
and a 'weak' one (which means 'independence of the subject'). Strong
objectivity seems to imply weak objectivity as its logical consequence
(i.e. as a necessary condition for it), but since the time men have
started to mistrust their possibility of really knowing the objects (and
this development may be analyzed by considering the history of
philosophy), they have been led to consider weak objectivity also as a
sufficient condition or, maybe more exactly, to confine to this the
whole meaning of objectivity.

All this is true and known, but there is a deeper sense, according to
which objectivity must in a way be related to the subject without
'depending on it', which deserves to be theoretically scrutinized even if
we leave out of consideration the historical links between the two
kinds of objectivity which we have just recalled. Such a viewpoint

seems to be of special interest if we are particularly concerned with 'scientific' objectivity, for an object is always for science not so much something that 'must be there' as something that 'must be known' and knowledge immediately and necessarily calls subjectivity into play, because it is always a product of the subject's knowing activity.

When a subject considers reality, he is inclined to think that all aspects of it that he perceives must be shared also by all the other fellow-subjects, because there seems to be no possibility that something real should not seem the same to one. But one is immediately led, through his commerce with the other subjects, to recognize that reality contains many aspects, many determinations, which appear to be agreed upon by all subjects, but it contains in addition several features that are real for me in this moment, but are not recognized as such by other people or even by me in other moments of my existence (e.g. dreams, perceptions had in a particular physical or mental state, etc.).

In such a way, a splitting takes place inside reality: there is a part of it which is real for all subjects (and we call it objective) and another which is real for me or for some subjects only (and we call it subjective). It is important to stress that no bipartition of 'reality' and 'appearance' is being proposed in such a way, because objectivity and subjectivity are both part of reality (for there is a difference, e.g. between having or not having a dream, which means that, as dream, it is real). On the other hand, it is quite clear that, of these two parts of reality, by far the most important turns out to be objectivity, not only because it is the only one to which we may make a constant reference when dealing with our fellow-subjects, but also because objectivity happens to be the only warranty we can advocate for reality when we try to establish it out of the short interval of our present experience (notice that every subject splits up, so to say, into a collection of different 'subjects' when experiencing the same reality at different times).

After having recognized the overall importance of objectivity, conceived exactly as intersubjectivity and so explicitly defined, one is immediately faced with an intriguing problem: knowledge as such is always a first-person activity, i.e. it is something which originates inside the subjects. How can it become 'intersubjective' or, as we could say, 'independent of the subjects'? At first one could think such a result to be obtainable in a completely satisfactory way only if every subject could experience other subjects' experience, i.e. to know their knowing, to perceive their perceptions, to be aware of their awareness, etc. But this is patently impossible and the fact that we actually agree with other men upon an enormous lot of characters of reality clearly indicates that the way to intersubjectivity does not require the impossible performance of looking into another person's mind. The reason why this has not to be the case is not difficult to uncover: in order for a notion to be considered objective, what must be evident to every subject is not the perception of such a notion by other subjects, but simply the *agreement* upon this notion among the different subjects and this agreement can be easily verified without entering other people's mind. It suffices to imagine how I could try to control whether a certain notion (e.g. the colour red) is *understood* (I do not say it is perceived) in the same way by me and by a generic interlocutor. In order to see that, I would try to find out a series of behaviours where the notion of red is involved (such as to press a red button, to select a red pencil out of differently coloured ones, to stop at a red light, etc.) and if I *ascertain* that the way my partner acts is exactly as I would have done in all those cases, I shall safely say that 'red' is an intersubjective notion between us, although I could still conceive the possibility that he might perceive the red subjectively in the manner in which I personally perceive the green. A quite similar discourse may patently be repeated also in the case of more abstract or sophisticated notions, like those of mathematics or physical sciences e.g. but we shall not go on discussing them here.

There is, on the contrary, another point which it seems necessary to stress, i.e. that the agreement upon the characters of reality, which yields objectivity in the sense of intersubjectivity, clearly appears as a consequence of performing some *operations* because, as we have noticed, it is not the way of *perceiving* a notion, but the way of *employing* it, that can make it objective, as we cannot have direct and personal knowledge of other people's perceptions, while we can have of their way of operating (the case of abstract notions does not constitute any difficulty, because language too may act as an operational tool, as it would not prove difficult to show).

It is scarcely worth noticing that the above maintained overall importance of operations and pragmatical devices does not imply any pragmatic conception of science, for the here stressed praxis is a strictly noetically oriented one: it is a praxis that yields knowledge and not a praxis which aims at transforming reality, although science may have subsequently also such a goal as one of its most direct consequences.

One has to be aware that in all that we have said science has been kept out of specific consideration, for we started with an analysis of the common sense idea of objectivity and went on to recognize some of its most peculiar features. It is therefore particularly interesting to remark that several points are nevertheless implicitly contained in this analysis, which are usually recognized as relevant in scientific knowledge.

First of all, the characterization of objectivity as intersubjectivity, which we found at the beginning to be one of the most widely accepted tools for making precise the notion of scientific objectivity, turned out to be perfectly justified inside our analysis. If we then consider what was said about the 'subjects' when speaking of the transition to the intersubjective knowledge, we must admit that such subjects were never conceived as conscious beings, as 'minds' or something of the kind, because we saw that consciousness, awareness, etc. are exactly such things that cannot be shared by a multiplicity of subjects, so that no reference to them may appear in the structure of intersubjectivity. Subjects, under such a respect, must simply be considered as 'recorders', as 'viewpoints' or 'generalized frames of reference' upon reality and what may be called objective appears hence as something that, thanks to the operations, may be held as *invariant* with respect to such different frames of reference. We have thus found the second of these characteristics which are most frequently advocated for scientific objectivity and, by the way, we met as well the correct way in which the notion of an 'observer' must be handled in science: such an observer is never to be considered as a consciousness, but simply as a 'generic frame of reference' or as a 'recorder', which may be equipped with more than one kind of instrument, apt to determine more parameters than the usual ones of space and time and also such to have 'interactions' with the objects it is supposed to investigate.

There are also other characteristics which are worth underlining: one is the fact that, if many subjects must be in the position of operationally controlling a feature in order it to be objective, it follows that scientific facts must at least in principle be *repeatable*. This is also a character of science that, provided it is carefully explained, cannot be denied by anybody and the careful explanation needed would simply amount to stressing that we never look in science for a repetition of an *individual*, of a singular fact, but of a generic fact of a certain *kind*. This means that also the so-called irrepeatable facts, such as stellar explosions e.g. are repeatable, not because this particular star which has exploded can explode again, but because the phenomenon of the stellar explosion as such must receive a scientific explanation inside astronomy, which states under which conditions it would be possible to observe it again. Needless to say, such a repetition has nothing to do with our possibilities of *producing* the fact again, but simply of *observing* it again, if certain precise circumstances are fulfilled.

We could say that, through repeatability we are able to come out of subjectivity and to broaden it to reaching the agreement of many subjects. We now shall consider a further characteristic, that is strictly related to this one but is taken in a slightly different sense and comes about when one considers that scientific objectivity ought to be something better than a pure 'broadened subjectivity'. This means that we should call intersubjective such features that are not simply agreed upon by *many* subjects, but that should be agreed upon by *all* subjects. But how to take the decisive step from the pure multiplicity to the totality? The answer is given by the general method by which in science one must try to give a foundation to statements concerning *all* elements of a too big or even infinite set: one tries to prove such a statement for a *generic* element of that set, which means that *whatever* element we take, it *must* satisfy the statement. In our case, we adopt the same procedure: if something must be valid for *all* subjects, we conclude that it *must* be valid for *every* subject and this can be expressed by the requirement that an objective property must be apt to undergo the *control* of every competent experimentator. The *controllability* (another basic requirement of scientific discourse) includes obviously repeatability and on top of it requires that the same results be always yielded when the same conditions are repeated.

From this viewpoint, one should be induced to give to *verification* as well more credit than it frequently receives: if one conceives verification in the restricted sense of supporting a *theoretical* hypothesis by finding out some empirically true consequences of it, one has obviously to recognize its shortcomings and admit e.g. the reasons that induced Popper to lay stress rather on falsification. But if one conceives verification in a broader sense, which includes also the fact of testing (by repetition of experiments) also the *empirical* statements of a science, one must admit that this part of verification is absolutely essential in order to give to scientific objectivity the character of some necessity and not of a purely contingent agreement among different subjects. Anyway, in order to avoid confusions or sterile polemics, we prefer to call *controllability* such a possible aspect of the meaning of verifiability.

After having proceeded to show how one reaches some of the most fundamental characters of scientific knowledge when considering how objectivity takes shape as an 'independence of the subject', it will prove interesting to see how the central features of the previous discourse are met symmetrically by going the opposite way back, i.e. starting from an analysis of objectivity as meaning 'what is characteristic of the (scientific) object'.

Our starting point will be, this time, the obvious remark that every science is not concerned with the whole of 'reality', but simply with some specific and restricted 'domain of objects', which fall under its competence. We shall now ask how the objects of a science can be given and it will be easy to see how the most intuitive answer to this question is completely misleading. In fact, one is instinctively inclined to consider the assignment of its fields of objects to every science as meaning a kind of partition of reality into sets of different 'things' (possibly with some overlappings), so that every science has its own slice of individual things about which it spends its inquiry efforts. How false this perspective is one easily recognizes when thinking how difficult it would prove to assign objects in such a way to the most part of sciences (already physics e.g. would be quite difficult to equip with its 'proper' objects). But the situation appears even clearer if one thinks that, actually, every individual 'thing' may become the 'object' of a great number of sciences, according to the viewpoint one takes in considering it. In such a way, what determines a scientific object turns out to be the set of viewpoints which characterize a certain science in its approach to reality. We have thus obtained again the distinction between 'objectivity' and 'reality', which we have already found in quite a different context, objectivity meaning now that part of reality which is being 'clipped out' by a science as constituting its specific domain of objects.

One can surely feel rather doubtful on seeing the different sciences being characterized by the fact of expressing some 'viewpoints' on reality, but such a psychological difficulty will be easily overcome if one considers what scientific knowledge turns out to be in the last analysis: it is simply a system of sentences. If we consider our problem from this position, it becomes simply the problem of how to recognize whether a given *sentence* does or does not belong to a certain science and this might open to us a useful way of treating the matter. Let us take e.g. the sentence 'it is warm in this room' and ask whether it belongs to physics. If one thought that it does because it speaks about heat, which is one of the recognized objects of physics, one would simply be contradicted by the fact that no physicist would accept to consider such a sentence as belonging to his science, while he could accept some seemingly very related sentence, like 'the temperature in this room is $35°$ centigrade $\pm \varepsilon^{0}$'. Why could the first sentence not be accepted as belonging to physics? Simply because a physicist has no means, as such, for assigning any truth value to it, while he can do this for the second sentence by resorting to a thermometer, which is one of the admitted instruments for testing sentences in physics.

A physician, on the contrary, could have accepted such a declaration made by a person in this room as a datum, and possibly as a symptom from which to start a diagnosis of his state of health. This example could be easily generalized. We could see that about every 'thing' one can formulate a lot of sentences, which may be classified as belonging to one science or to another according to the 'criteria' which may be applied for establishing their truth value. We can in such a way reach the conclusion that what really determines the domain of objects of a science is every time a certain set of 'criteria', which allow one to assign an immediate truth value to some sentences. It is thanks to such criteria that a science can 'clip out' its objects from reality, by applying them to every 'thing' with which they are confronted.

If we now try to understand better what such 'criteria' really are, we easily see that they are of the kind of taking some instruments and employing them according to some instructions (physics, astronomy, microbiology, etc.), resorting to reagents, electric currents, heating procedures, etc. (chemistry), looking for documents of different kinds and putting them to some standard scrutiny for establishing reliability (historiography), taking texts, manuscripts and similar things and handling them according to some recognized procedures of collation (philology), and so on. In all these cases we can always see applied, with a sensible flexibility and a sufficient generality of meaning, the same method of using *operational* procedures, which implies the resorting to some instruments handled according to some prescriptions or rules.

One can thus draw the conclusion that the above mentioned 'criteria' are operational in character and they serve to 'give', to 'shape' the objects of every science (as they concretize the 'viewpoints' under which 'things' are considered in order to become scientific 'objects') providing at the same time the 'data' of every science (which are but the 'immediately true' sentences which must be recognized as such by applying those criteria).

A deeper scrutiny, which we cannot carry out here, would show that these criteria 'operationally define' certain basic predicates of a science (which could be labelled as observational or empirical, but even better as 'operational') and the 'object' must be conceived as a relational structure established between all these predicates. Further theoretical predicates, which can and must surely enter into every science, are supposed to be concerned with its objects only if they turn out to be logically connected with some totally operational sentence which concerns, *thanks to that*, the objects (and here we find again the notions of verifiability and falsifiability involved in a deeper sense).

As one sees, we reached along this new path the same conclusion of the operational feature of scientific objectivity, that we had already recognized before and this independent confirmation is surely very significant for the correctness of the perspective we have adopted here.

The least which one would expect after that conclusion would be a declaration of full devotion to that epistemological position which is known as 'operationalism', but the question is not so easy. We surely pointed out some reasons for which no scientist can deny to be in a way an operationalist, but the actual operationalists (like Bridgman and his followers) arrived at some conclusions which do not seem quite acceptable, by maintaining e.g. that all concepts in science are operational (while theoretical concepts cannot properly be conceived as such). They are on top quite unaware of the role played by operations in securing objectivity in science, so that Bridgman could take some well known positions of an extreme subjectivism about science in general.

It would now be quite trivial to verify that other characters, such as those of repeatability, controllability, invariance, are elementary consequences of having conceived scientific objects as constructed by operations and that, last but not least, the fact that instruments and rules for applying them must be taken as known, as granted and unquestionable in order for somebody to enter the field of a certain science, yields also the result that operational scientific objectivity is as well an objectivity in the sense of intersubjectivity. In such a way, as we have announced, the two symmetric ways of defining scientific objectivity lead to a complete overlapping of their respective basic characters so that we could be entitled to repeat, in a way, what Kant said of his transcendental conception of knowledge in general: the conditions for the possibility of (scientific) knowledge are the same as the conditions for the possibility of the objects of this knowledge.

Let us now enter a final stage of our inquiry and reconsider the general opinion with which we started our reflections, i.e. that science is no longer concerned with truth, but simply with objectivity. We shall now devote some analysis to the relations existing between the concept of truth and that of objectivity as it was characterized here.

We shall first of all distinguish between a substantial and an adjectival use of the truth concept: according to the first, truth constitutes a kind of autonomous and immutable realm, which can be uncovered step by step like an unexplored continent (in such a sense one can speak e.g. of the truths of physics, the truths of Christian faith, etc.); the adjectival use of truth, on the other hand, conceives it as a property which can or cannot be attributed to *sentences*, according to some conditions which have to be specified. Having considered science as a system of sentences, we shall confine ourselves only to the adjectival use of the truth concept. Now, when must we say that a sentence is true? The ancient answer given by Aristotle and perfected in the formal definition of Tarski may be summarized as follows: 'true is a sentence which tells what there is; false is a sentence which tells what there is not' and as the definition of truth relies here upon the reasons of 'being' and 'not being' one cannot avoid recognizing that every sentence cannot be true *and* false at the same time, but it must always be *either* true *or* false.

As a first conclusion we must say that, if every science has its knowledge expressed by declarative sentences, these must be either true or false and, on the other side, one does not see how scientific utterances, which are surely not imperatives, nor questions, nor exclamations, nor deontological prescriptions, etc. could be anything but declarative sentences. A certain difficulty in admitting that may come from the fact that when speaking of the truth of sentences one usually forgets that a sentence is only true or false *about* some specific subject matter. In such a way, every scientific sentence is to be either true or false *about the objects* of this science and not just generically or, so to say, about *things*! Once this is well understood, one easily recognizes that the aim of every science is to build up a system of *true* sentences concerning its own *objects*, so that objective knowledge aims at being also a true knowledge.

Some of such sentences are surely available, and they are the empirical data which come out automatically when one applies the basic operational criteria and in such a way obtains at the same time the objects and the first empirical *true* sentences concerning them. All the other sentences which are non-empirical, but still speak about the objects, must also be either true or false, because they cannot help telling about these objects either what there is (and they are true), or what there is not (and they are false). Everything that has been said about the impossibility of acquiring definitive truth in science, or about the inadequacy to this end of any big number of confirmations etc., must surely be considered as well founded, but it does not concern *truth* proper. A sentence, once it is granted that it speaks about something, cannot help being either true or false. What is not granted, on the contrary, is our possibility of establishing with *certainty* which of these two cases does actually occur for every sentence.

And now an engaging conclusion comes out here: if every science is made out of sentences and *all* which are accepted in it are supposed to be true and *most* of them are actually true, it follows that such true sentences 'tell what there is' and, therefore, they are ontologically committed or, if we prefer, the objects about which science speaks have an ontological existence. If they had no such one, it would simply follow that everything is said about them 'tells what there is not' and therefore it would yield a false sentence. In such a way we recognized that every science must admit an ontological support of its discourse, if it is not ready to admit that all its sentences are inevitably false.

If one understands these facts, one does not feel any difficulty in admitting the existence also of entities corresponding to theoretical concepts or constructs (like electrons, Super-Ego, etc.) because the only way of denying an ontological stature to such entities would be that of proving false every sentence concerning them in the respective science.

But, once this is said, one must never forget that such entities have an ontological existence as 'objects' and not as 'things' of everyday experience. If one forgets that, one needs e.g. such strange features as 'complementarity principle' in quantum mechanics, which simply tries to make palatable to our poor intuition something which is difficult to admit only if we think of waves and corpuscles as 'things' of our intuitive world picture, while they need not come to a collision if they are thought of as non-intuitive 'objects' of their proper science. If one understands, in other words, that the ontological character is not bound to a palpable or intuitive perception of what exists, but simply to the fact of existing, one should not have any troubles in recognizing the ontological commitment of science.

We are maintaining in such a way a position which is neither a rough materialism nor a rough idealism. It is not rough materialism because, in admitting that scientific knowledge never has to do with 'things' proper, but with objects, which are obtained when things are submitted to some operational manipulations, it recognizes in a way the fact that no conflict exists between scientific knowledge and scientific data, for the same operations yield at the same time the data (which are 'constructs') and the noetical knowledge of them. On the other hand, it is not a pure idealistic conception, because it does not conceive objects as a pure projection or creation of mind in its knowing activity. In fact, the objects are surely 'constructs', but 'operational' and not 'mental' constructs, what even implies that they must come out of a structure of material and preexisting 'reality', which is given us in everyday experience or, as Husserl would say, in the *Lebenswelt*.

Science, in other words, does not grow up in a vacuum, but every knowledge always proceeds out of a world of 'things' and of preexisting knowledge, which constitute in a way its body of 'material' conditions to proceed to its own edification. What is here kept of idealism is the negation, *inside knowledge*, of any conflict of thought and reality, while what is not accepted is the hierarchical dependence which makes reality a mere production of thought. What is kept of materialism (or, to say it better, of 'realism') is the idea of a reality independent of every *singular* noetic enterprise and giving the conditions for it to start, while it is admitted, against the most current realism, that knowledge can only take place as far as reality is structured in constructs, which are the only proper objects of scientific knowledge and inquiry.

Such a form of what might be called an 'objective realism' seems to me the best suited to act as a foundation for a faithful account of scientific knowledge, because it has not that naive substantialistic feature that would coincide with a gnoseologistic dualism (we say e.g. that an object *is* the totality of its related properties, and not that it *has* them); on the other hand, it does not go so far as to resolving completely the 'substance' into the 'functions' (to employ Cassirer's well known opposition). We could perhaps say that it rather conceives the substance as a structure, and this allows us to say e.g. that an object *is* the structured totality of its properties but also that it *has* every one of them as well, and at the same time this conception does not consider the structure as made out of pure relations or functions, because it includes separate properties which enter in the relations and can meaningfully be considered in themselves as well.

It should also be apparent from what was said here that such a conception is general enough to prove suitable for sciences which have to be considered empirical without needing to be, on the other hand, only 'natural' sciences.

University of Genova

GENEALOGY OF SCIENCE AND THEORY OF KNOWLEDGE

More recent philosophies of science have often tried to create a model of science and then compare philosophical conclusions with concrete research and discoveries. For instance, one could start with the assumption that (natural) science is essentially based on empirical facts or sense-data and that, consequently, all theoretical concepts or laws have to be reduced (by some correspondence rules) to a primitive empirical language or observation. Any model is useful insofar as, in accordance with some well established principles, it produces some valuable results. The difficulty with the empiricist or quasi-empirical approaches has been precisely the failure to produce any important discovery in spite of well argued guide lines; and what is still worse, some concrete examples or illustrations of these philosophies often showed misunderstanding of actual research in physics, biology, and other so-called empirical sciences.

My approach is a quite different one. To understand the range of human thinking, or relationship between man and the cosmos, we have to first examine the real sciences – how they developed in past history, certainly not in every detail but in the main trends (if such trends exist at all). It may be presumptive to suppose that we can understand axiomatic systems (or even some 'so-called' fact-statements) without grasping the whole framework of concepts and practical procedures used in any particular discipline. Every student of mechanics knows how long he has to practice the 'space-time-force' relations in mathematical and experimental ways before he is able to apply and fully understand Newton's law of motion; and these relations are deeply connected with the movements of his body and with his seeing. Without such experience the best logician will not be able to do anything with a formal system, and the fruitlessness of logistic attempts in contemporary physics is a strong argument for a differing point of departure.

Butts and Hintikka (eds.), Historical and Philosophical Dimensions of Logic, Methodology and Philosophy of Science, 173–183.

Certainly many philosophers might object (and some actually did so) that the whole of science is just a conjecture (ein Europäischer Zufall) and I would agree that much research has been unnecessary, fortuitous, conditioned by personal taste and dominant views. Nevertheless, when we look into such different cultures as West European, Chinese, and African, they have not produced basically different physics, although the original languages, habits, and concepts have been very different. The second striking argument is that there have been many simultaneous, independent discoveries. So we can safely conclude that the main trends of development of geometry, mechanics, optics, and thermodynamics have been necessary – taking into account the activities in a more developed society with its architecture, geodesy, use of tools and machines, traffic systems, vapour-engines, etc. It is obvious that these activities are deeply rooted in the structure of the human body and basic human needs.

I am not taking the *a priori* position that there are some independent entities, properties and laws which are discovered in time; but the origin of science suggests that the basic interaction between men and environments is very similar for all human beings due to the same, or at least, very similar, biological constitutions and the similar basic work and personal relations. Such a universality presupposes neither a cosmos with inherent properties, nor a subject (or Kant's reason) with definite categories or forms of intuition, and it is related to the basic interaction between men and nature, interaction historically developing and never reaching a final goal. The assumption of the realm of universals (whether ideal or material) would put any theory of knowledge in an intolerable situation: How can we grasp such absolute entities or laws, and moreover, how can we explain the variety and history of human concepts?

Taking no refuge in the Platonic realm I am at the same time refuting the arbitrariness that has been advanced by many investigators of different societies. There is no need to take a middle, inconsistent, eclectic way between absolutism and relativism. If one assumes that all human concepts and languages are determined by social forms he cannot avoid questioning how these forms have been created, and then he has to assume either arbitrariness or more general causes and we are in the same position as before. In this paper I am not examining all factors relevant to human history – manners of material production, or the role of different social structures and institutions – my concentration here is primarily oriented towards the genealogy of natural science.

Newton could not formulate 'Philosophiae Naturalis Principia Mathematica' without using Euclidean (or similar) geometry. It would be very difficult, if not impossible, to define some geometrical concepts and procedures (for instance congruence) without the use of motion or some idealised features of solid bodies or rays of light. The geometrical system and mechanical system are closely linked, and in Einstein's general theory of relativity became united in a field theory. Certainly, the field concept of Einstein, as well of Boskovic and Faraday, is rooted in the concept of Newton's force. Kirchhoff has accused Newton of anthropomorphism but without human elements no advancement of our knowledge of nature would be possible. Space, time, and force (linked with cause and effect) are basic to all human existence and interaction with the environment. We produce the changes in the world around us and we are accustomed to look for a cause for any change in space and time. Physics is primarily a very complex and interwoven set of actions, thoughts, and procedures which can be, when concluded for some sections and structures, put in a stenogram of laws, understandable to students of relevant disciplines. Such axioms can be expressed in logical forms of 'is' or 'are' sentences; and this shortened way of expression hides the character of human action with its intentions, evaluations, biological and historical conditions.

Principles of geometry and mechanics taken together constitute the framework of quantitative measurements, expressed in primary units of length (meter), time (second), and force (gram). Physicists build their apparatuses according to acquired skill, some general laws and relevant concepts about interaction between apparatus and introduced object or process. From this standpoint it would be senseless to ask whether these principles are purely 'objective' or 'subjective'; they will be understood only in inseparable relations between men and nature. Consequently the separate thing and property, so basic for Russell and logical positivism, would lose their meaning. To say that 'the chalk is yellow' represents a simple, basic statement or observation is really neglecting how complicated and highly specialized color differentiation is for humans.

Geometrical and mechanical systems of axioms are so formulated as to contain certain laws of conservation or invariant principles; for instance, laws referring to quantities of energy and momentum. And now comes a surprise. These laws of conservation, so well established by many measurements of macroscopic bodies and processes, fail in special, characteristic, and repeated circumstances. The scientists faced a dilemma: to reject the universal validity of these principles, or to assume the existence of invisible fields and elementary particles which could absorb and emit momentum and energy. For the strict empiricists only the first solution was possible, but the physicists convinced of the validity of the laws of conservation really proceeded in the opposite direction. I am as sure that an atom exists as I am sure that my boat will move when I jump into it. Due to the discovery of the invariance principles and symmetry properties we are able to transcend the immediate, macroscopic reality which encompasses our body as one of its parts.

Physical apparatuses, like cloud chambers or G.M.-counters, detect macroscopic effects of electrons or protons, but the 'causes' are invisible and untouchable by our sense organs. Quantum theory is primarily concentrated in the transformation of this underlying reality into our macroscopic framework. According to Bohr's principle of correspondence we have to ascribe to atomic systems or even particles some of the quantities of macroscopic bodies, like energy, momentum, position in space and time, but these quantities have to be applied with particular care or limitation as provided by Heisenberg's uncertainty principle, to take into account paradoxical appearances of corpuscles and waves and of quantum jumps. When Planck's constant becomes negligible, Quantum Mechanics has to, according to Bohr, transfer into classical mechanics or – to put it in less strong form – some parts of quantum mechanics have to do that, but not necessarily *all* parts which make a consistent whole – an essential amendment not understood by many analysts.

This transfer the Q.Th. to Cl.M. is not unique; there may be different possibilities. What we know at first is only the final result, classical physics, and it has been guessed by Heisenberg how the classical concepts and laws have to be expanded and transformed to get a consistent mathematical scheme for atomic processes. It is obvious that matrix mechanics, like its mathematical equivalent – wave mechanics – has a physical meaning only in relation to macroscopic physics. But this is not a formal, logical, deductive relation; classical physics is not a limit case of quantum theory for $h \to 0$. Any attempt to establish an atomic theory independent of classical physics, or macroscopic concepts, would inevitably end in incomprehensibility. On the other side, the ascribed atomic quantities must not be understood and treated as usual macroscopic quantities, and they cannot be integrated in an objective process in space and time; their physical meaning is contained in transformation into macroscopic effects; there is no sense in asking what properties an electron has *before* an experiment. Physics is essentially connected with measurement and reduction to a meter-second-gram system, and it is hardly meaningful to speak about a physical reality beyond this well-established physical framework.

Under the genealogy of natural science I do not understand the full history of scientific research with all its actual conjectures, presuppositions and failures, but only the main trends which are closely linked by consistent inferences and procedures, so that newer branches are only meaningful in relation to older ones. This genealogy starts with Euclid's geometry, Newton's mechanics and geometrical optics containing essential human interaction with nature as reduced to the basic concept of space, time, and force. Once we had this basis we could develop thermodynamics, particularly the first principle of conservation of energy, and afterwards a more accurate definition of temperature (the second law). After mechanics, especially the mechanics of gases (pressure-volume law),. and thermodynamics had been established modern chemistry could begin. At the same time Newton's force had been by Faraday transferred to the concept of the field. After differential equations of the electromagnetic field had been discovered by Maxwell, and after optics had become a part of electrodynamics, the main problem was how this new field was to be united with classical mechanics, and this task was accomplished by Einstein by radical revisions of the traditional concepts of space and time. H. A. Lorentz tried in vain to construct a consistent system of all classical physics based on electrons; such a consistent theory could never be reached, and with Planck and Bohr there started a break-through of a quite different concept of reality. Without pretention of completeness we can represent the development and interdependence of these various main branches of science as a descending tree.

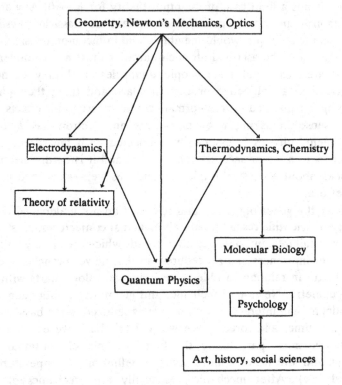

It would be wrong to say that Einstein overthrows the classical theories of Newton, Bohr and Heisenberg, and so on. We could never understand relativity theory without first grasping how, with Einstein, different elements of the electromagnetic field and mechanical motions have been brought together into a system of axioms: relativity of inertial systems and constancy of velocity of light; and the same is true about the Quantum theory (as has been previously explained). How could anybody affirm that the scientists who adhere to one set of axioms or paradigms do not understand the scientist who sticks to another one? Or pretend that they speak each to the other without finding a link or bridge? The idea of scientific revolution is very attractive when one views the whole change of concepts in the 20th century, and this representation has been used by many scientists, but this has often been misleading, for instance in picturing that the old physics had been abolished and a new one erected on virgin soil. On the other hand, the new development did not leave the old, classical basis untouched; everybody who now writes a textbook on Newtonian mechanics refor- mulates the basic concepts, definitions, axioms, and equations in a way that will make a more natural transition to the theory of relativity and Quantum mechanics; and while invariance principles are hardly stressed in Hetz's excellent exposition of mechanics, a contemporary theoretician would bring these principles to the forefront (and the importance of the measurement as well). So we cannot regard any of the classical branches of mechanics as completely finished and closed. All these different branches constitute one living tree with characteris- tic relations between themselves and deep roots into our human existence and daily experience.

Science is not a simple growth of knowledge or accumulation of facts, but has a characteristic structure; the relations between two close branches or disciplines are not of purely logical form but are subjected to evolution in the sense that in the first branch a change of concepts and procedures takes place when finally the second branch is full- fledged with its own concepts and laws – as we saw for instance, in the development of classical electrodynamics from Newton's mechanics, or Bohr's model of atoms from the structure of solar system. This transition from one branch to another is not just a fortuitous guess but contains many mathematical formalities and experimental procedures. It looks at the end as if no other possibility exists for a new theory. Science is a very unique enterprise of investigators, and a demarcation line between science and non-science can be drawn with regard to genealogy of science. If some kind of effort, like magic or astrology, cannot be included in this ramifying tree, then I shall not regard this as a scientific effort. The proof of the burden that another genealogy of science is possible, or even desirable, lies on philosophers who are trying to create a perfect model of empirical science.

Each branch in our genealogy of science contains both empirical and theoretical concepts, and so relations between different branches can- not be described as inferences from observation to hypothesis. For instance, Bohr's principle of correspondence has little in common with the so-called correspondence rules of logical positivism which would establish a bridge between hard facts (sense-data) and theoretical concepts; or in more sophisticated form, between a primitive (empirical) and more complex (theoretical) language. When we transfer quantum mechanics into classical mechanics, we may in fact translate something more theoretical into something less abstract, but classical physics also includes theoretical concepts and practical procedures. A separation of theory from experiment, or of thinking from practice, would make human actions incomprehensible; and science is precisely a very elaborated human enterprise. Could I say that the length of my room and the hours of my walk are something simply observational, or to the contrary that they are only forms of my intuition? The insoluble philosophical problems arise from an absolute separation of subject and object.

Genealogy of science confronts us with many problems which are traditionally regarded as purely philosophical. First of all, this structure of science conspicuously contradicts the empiricists' uniformity or flood of impressions, sense data, and observations. When I see my friend Martin standing before my house does this observation mean the same as the concept of an inside and an outside of a box? Some space relations are deeply rooted in all our observations, whereas others may just be fortuitous and framed by more general and even existential relations. Secondly, could one really believe that such an inductive generalization as 'All ravens are black' has the same fundamental significance as Newton's law of motion? Would a satisfactory explanation of 'why this bird is black' be equivalent to applying the above stated law? The biologist would probably undertake genetic, physiological, and chemical explorations to answer this question. The same grammatical and logical form has confused the very character of physical laws.

The basic problem of theory of knowledge has often been seen in relations between our concepts (thoughts) and outside reality. So stated, the problem is confusing, if not insoluble. Starting from such a position, Boltzmann would see in the development of physics a more and more accurate picture of outside things (a position taken over by Lenin and dialectical materialism), and Russell would initially, like Wittgenstein in the *Tractatus*, for any logical subject, look for a real thing, fact or state of affairs. The original confusion is in the concept of an outside, independent, objective thing. The outside of a box is quite a reasonable concept, and with some care one can speak about bodies outside or around his body; but *the physical reality outside of our mind is a very odd concept*. The physical concepts and laws, us we have understood from the genealogy, are rooted in human action and are related to the interaction between men (society) and nature (environment).

If I walk in a meadow and suddenly exclaim, 'this rose is red', I would be very surprised if my escort would ask me, 'Is that true'? If I simply exclaimed, 'red rose', there would be no logical question for my escort; but I could mean the same. It is obvious that the question of truth does not arise at that level. Limiting ourselves to such 'is' statements and characterizing them as being true or false (or neutral) would not only cripple human language, but would put the truth on too low a level, if indeed it would not end in incomprehensibility. We cannot find truth apart from human life and the scientific enterprise. 'The red rose' is something which we learned to understand in the long human evolution and our personal education. If Martin numbers his steps and makes the statement, 'It is seven', and all of us other than him find that he executed eight steps, I would say, 'Martin, that is not true.' Somebody has to undertake an exploration or to make a claim before the question of the truth becomes acute, and then we shall find the truth in the coherence or consistency of all his statements, predictions, procedures, and applications. The theory of knowledge has to analyze the more stable, lasting, general concepts, and most often used procedures (or things) without a too hasty presupposition about their separate existence. Human acts are very complex, with never fully determined motivations and intentions, in an historical, social, and natural environment, and it looks hopeless and meaningless to isolate and absolutize a particular entity. The assumed existence of some individuals with inherent properties would bring our understanding to an impasse; but on the other hand, to look at everything through some general categories (as Hegel did) would lead to the opposite, equally incomprehensible extreme.

The genealogy of science throws some light on the old mind-body problem. Since Descartes defined the independent substance of matter by extension in space, and so dehumanized the concept of space, the mind appeared as a ghost 'inside' of the head. A similar difficulty remained for Dubois-Reymond who with good reasons assumed that mental processes are connected with physiological states of the brain; if the basic atomic movements represent a determined and closed system, can spirit be more than a mere shadow of physical processes? As long as we stick to the premise of Descartes (and classical mechanics), there is no way out of this difficulty. But we have to look into the whole genealogy of science instead of rushing to any reduction procedure. Prima facie, the beginning of science is clearly rooted in our body activity; with the Quantum theory we transcend the primitive concept of the body. Certainly, there are in every living organism some definite, genetically developed macro- or semi-macrostructures; nevertheless, the basic processes are micro fine and cannot be fully transferred into a macroscopic framework. Few electrons or quantum jumps provoke big macroscopic effects or even the death of a living being. Consequently we lose the concept of 'body' at the lowest levels of biological existence.

The living organism has sometimes been compared with a physical device with great potential of amplification and self-regulation. This comparison is still misleading if we take into consideration that in a physical apparatus we can separate macroscopic structure and function built upon some physical laws; whereas such a separation between *macro* and *micro* is impossible in a living organism. We cannot look for some electronic impulses or quantum jumps producing certain visible effects in a defined macroscopic framework, as we do in a cloud chamber. Such a definite framework does not exist in any living being. The simple reduction of biology to physics is a gross misunderstanding of the range of physics. The 'body' disappears at the lowest levels of life. We have to use different approaches to understand and control what happens there. The Quantum theory did not establish 'the freedom' of the human being, but it has eliminated crude determinism which has been, in the past, the main obstacle for more complex study and more harmonious relations between natural sciences and humanism.

Although classical physics is closely related to macroenvironment, I would not say that our concepts, even in Euclid's geometry and Newton's mechanics, are absorbed in interactions between our body and external bodies. Human thinking always contains something more than direct contacts with events and facts; we have the power of generalization and imagination which are able to go to the utmost limit and create different worlds. The transcendent element has been, from the beginning, present in human thought, and that made possible the transcendence beyond the immediate reality of our daily experience and senseperception. The genealogy of natural science shows as well the possibilities of human understanding as our existential position in the cosmos, a unique human undertaking whose final conclusion none can grasp.

University of Zagreb

II

HISTORICAL PERSPECTIVES
ON THE CONCEPT OF MATTER

B. M. KEDROV

EVOLUTION OF THE CONCEPT OF MATTER IN SCIENCE AND PHILOSOPHY

1. The concept of matter is one of the fundamental concepts of scientific and philosophical knowledge.

Like other scientific concepts, it developed through a long history during which it was subject to growth, refinement and revision reflecting at every historic epoch the general level of scientific knowledge of the world that had been reached at a given stage of development of human thought. The higher the level was, the more meaningful was the concept of matter itself, even if it were not formulated by means of a more or less extended definition, nor had an appropriate term. Nevertheless, whenever the fundamental elements of the universe were mentioned, one essentially implied what was later called 'matter'.

Since the concept of matter developed in analogy with any truly scientific concept, it passed through definite stages corresponding to those passed through by human knowledge in general. It is known that Hegel's 'Theory of knowledge' (Cognition) proceeds in categories of individuality (I), peculiarity (P) and universality (U). This theory has a profound rational meaning. The knowledge of some new class of phenomena begins with the statement of individual, discrete facts as yet not related to each other (I). Then during the search for mutual relations hidden in them the knowledge raises the unitary from its U and transfers it in P through the primary classification of facts, for example, according to tests for identity and difference. Such organization opens up a possibility to find a universal, regular connection between the already established groups of facts, to discover a law for a given circle of phenomena and thus raise the peculiar from its P and transfer it into U.

This is how in the general case the regular process of development of knowledge from I through P towards U occurs. In its own peculiar way it was exactly this type of a regular cognitive process that took place in

Butts and Hintikka (eds.), Historical and Philosophical Dimensions of Logic, Methodology and Philosophy of Science, 187–208.

the course of development of the concept of matter. This was accompanied by constant interweaving either of its first two stages (*I* and *P*), or the higher stages (*P* and *U*) or of all three simultaneously.

I. ANCIENT SCIENCE

2. The problem of the fundamental elements of the universe was already formulated in The Miletus school of ancient Greek philosophy. Here what corresponded to the concept of matter appeared in the form of a concrete body: in Thales' theory it was water, in Anaximenes' – air, in Anaximander's – a certain indefinite fundamental element (but exactly as a single body) – apeiron. This was the stage *I*; it was not completed at the early times of the ancient Greek philosophy but stretched its filaments far in the future, in the New time. Thus, Thales' idea of water as the protomatter is related to Boyle's concept of water being the primary matter (the 17th century); at the beginning of the 19th century the influence is still seen in Prout who interpreted hydrogen as the prototype implying that all chemical elements are formed of it, and still further, towards the 20th century when the nucleus of hydrogen that constitutes a component of nuclei of all other elements was named the proton.

In exactly the same way Anaximenes' idea of air as the fundamental element of the universe is related to the subsequent discovery of oxygen in the 18th century, which for Lavoisier was a fundamental element representing the basis of all of chemistry. Later in the 19th century when after the Boyle-Mariotte law other laws of ideal gases were discovered, it was exactly on the material of the rarefied gases that the thermodynamics and the molecular kinetic theory, and then the theory of diluted aqueous solutions (falling in the sphere of applicability of the above laws) were developed. Therefore, gases ('air') being the simplest physical systems served as the starting point for the development of molecular physics and physical chemistry. Thus, in the 19th and the 20th centuries there was a continuation of the development of ideas the sources of which belonged to deep antiquity when the concept of matter had not yet left the frame of *I*.

3. But already in the ancient Greek philosophy there appeared such concepts of matter that actually took this concept out of the initial narrow limits *I* and raised it to the level *P*. The development of the concept of matter proceeded in this direction following two different paths: the first implied a separation of a special group of fundamental elements containing the above mentioned water and air complemented by earth and fire. Thus the representative of primitive natural scientists Empedocles formulated a theory of four 'roots'. Various combinations of these four 'roots' lead to all objects of nature. Later, beginning with the 19th century, when organic chemistry appeared, it was in a way similar to the ideas of Empedocles in the sense that a practically infinite number of organic substances results from various combinations of four basic elements (C, H, O, and N).

Further evolution of the concept of matter in the frame *P* along this path has led to Aristotle's conception of four elements capable of mutual transformations. This special capability as well as other special properties of these elements homeomeria was derived by Aristotle from combining various elementary properties (warm or cold, dry or wet). Thus, Aristotle's concept of matter was actually based on *P*.

The other path of development of this concept led at the same time to the atomic theory as a special structure of matter. Through the conception of another primitive natural scientist Anaxagoras concerning qualitatively different homeomeria it resulted in Leucippus' and Democritus' idea of quality-free atoms (indivisible) differing only in magnitude, shape and position, and also in weight (heavy or light atoms). Later this conception was developed by Epicurus and Lucretius. Thus, the atomic theory as well the concept of matter was based on *P* (the discrete nature of matter, the quality-free property of the indivisibles).

4. But already in the atomists there appeared outlines (although quite indistinct) of the concept of matter, constructed on the basis of *U*. For example, in his poem 'On the Nature of Things' Lucretius wrote: " ... res quoniam docui non posse creari de nilo neque item genitas ad nil revocari." This expressed a principle: nothing from nothing and nothing into nothing. In such general form the principle of conservation of the basis of the whole reality appeared and as such basis the matter in its P was actually adopted, since the author discussed things in general, regardless of their properties, composition and structure, in other words, regardless of their unitary and peculiar properties.

The principle of conservation of matter expressed as U in this abstract, natural scientific form allowed formulation of a concept of matter itself on the basis of U also in an abstract, natural scientific form. Later this principle as a general principle was subject to concretization when the author implied, instead of the conservation of the basis of things in general, the conservation of certain properties of matter, first the qualitative ones, and then the quantitatively measurable. Thus, in the 18th century Lomonosov and Lavoisier discovered the law of conservation of mass of substances in chemical interaction and Lavoisier directly referred to the principle 'nothing from nothing, nothing into nothing'. This was also an example of influence stretching from the ancient philosophers to the scientists of the New Time.

The development of the concept of matter in the ancient school took place on the whole in an atmosphere of undifferentiated science under the auspices of philosophy, which was still sufficiently indeterminate and essentially natural philosophy. Nevertheless, such an underdeveloped, diffuse and therefore naive approach already clearly showed a trend towards the transition of the as yet unshaped concept of matter from I into P and further from P into U. This trend contained the embryo of the whole further course of development of the concept of matter in science and philosophy, beginning from the 17th century and ending in the 20th century.

II. SCIENCE AND PHILOSOPHY OF THE NEW TIME
(XVII–XIX CENTURIES)

5. In the time of the Renaissance we observe an ever accelerating process of differentiation of sciences on the basis of mass penetration into them of the analytic method of knowledge. The formerly single science branches into independent disciplines, first the mathematical sciences, later the natural sciences, beginning with physics and chemistry and ending with anthropology, and finally, the social sciences. In the course of the struggle during the 17–18th centuries between materialism and idealism the concept of matter as a philosophical concept evolves. It is associated, however, with analytical properties related to scientific theories of matter, which were formulated in the sphere of mechanics, physics and chemistry, i.e. in the sphere of inorganic science. All versions of the concept of matter of this kind are based on P, the P itself appearing as a result of a purely analytical approach to the investigation of the objects of nature. The two paths that came into being already in ancient times were expanded in the 17th century with a third one according to which matter is defined qualitatively as mass. Such a concept of matter is presented in Galilean mechanics and in a detailed form it appears in Newton's 'Philosophiae naturalis principia mathematica'. A special mechanical property of macroscopic bodies (the mass) becomes the defining property of matter. In connection with this, weight appears on the scene, as a characteristic of the material property of a body, since the mass expresses itself through weight. Accordingly, in the 18th century the quantitative law of conservation of matter is formulated as the law of conservation of mass, or weight of bodies (Lomonosov and Lavoisier).

In chemistry the analytical method leads to the definition of a chemical element as the limit of the chemical decomposition of a substance (Boyle, Lavoisier, and Dalton). Therefore matter is regarded as the set of chemical elements. In the atomic concept of Boyle, Lomonosov, and Dalton matter is interpreted as the set of indivisible atoms; in Dalton's theory different kinds of atoms correspond to different kinds of chemical elements. All such interpretations of the concept of matter are based entirely on P since their background is formed by either special properties (the property of being a chemical element) or special structure (atomistic discreteness) of the bodies of nature.

D. I. Mendeleev advances the concept of substance with its characteristic of ponderability as the equivalent one to the concept of matter (see Mendeleev, 1905, p. 1). In his concept of substance Mendeleev unifies the former scientific versions of the concept of matter, based on P (through characteristics: mass, or weight; the property of being chemically non-decomposable, or elementary; the atomistic property, or that of being constructed of indivisible, 'last' particles). According to Mendeleev, "Substance or matter is what, filling the space, possesses weight, in other words, represents masses ..., what constitutes the bodies of nature and what is subject to displacements and natural phenomena".

6. Along with the scientific concepts of matter, based entirely on P and related to the analytic approach to the study of nature, we observe in philosophy the gnosiological concept of matter characterized already by the features of U and following from the opposition between matter and consciousness, reality and spirit, the physical and the psychic. The 18th century idealists either deny the reality of matter thus rejecting the very concept of it, or deny its cognizability or interpret it as something secondary with respect to forces as something primary (the dynamic theory of Kant supported later by Shelling and Hegel).

Matter is treated by them as something inert, deprived of the inner activity, or self-motion. The materialists of the 18th century, the propounders of mechanism, while recognizing the primary nature of matter, identify it with the mechanical mass or the set of individual atoms. This means that they leave this concept at the level P and do not raise it on the whole to the level U except in those cases when they formulate the question of the theory of knowledge (e.g. Holbach).

The philosophical concept of matter as an object of knowledge was advanced also by Feuerbach in the 19th century. In his conception, however, the material aspect U has not been supported by concrete historical arguments and was rather abstract. Feuerbach essentially reduces the issue to opposing matter to spirit, the physical to the psychic, and the objective is interpreted as what exists outside of us.

7. The concept of matter in its entire volume, as a philosophical one including synthetically all special manifestations of matter and therefore constructed on the basis of U, appeared in the works of Marx and Engels. Engels in his 'Ludwig Feuerbach and the End of Classical German Philosophy' (1886) has put the concept of matter on a general foundation in the form of the materialistic solution of the fundamental question of every philosophy. As a result of this the concept of matter has appeared as a conceptual expression of the gnosiological relation which constitutes 'the great basic question of all philosophy, especially of more recent philosophy' – 'the question of the relation of thinking to being, the relation of the spirit to nature – the paramount question of the whole of philosophy' (Marx and Engels, 1973, pp. 345–346).

As is well known, this issue has two aspects: (a) which is primary, spirit or nature – did God create the world or has the world been in existence eternally? (b) in what relation do our thoughts about the world surrounding us stand to this world itself? Is our thinking capable of the cognition of the real world? Are we able in our ideas and notions of the real world to produce a correct reflexion of reality? In philosophical language this question is called the question of the identity of thinking and being.

A consistent materialistic solution of the fundamental problem of philosophy is characterized by two aspects: (a) recognition of the fact that the matter, the being, the reality is primary while the consciousness, thought, spirit is secondary, and (b) the affirmative answer to the second question formulated above. Hence the concept of matter itself the essential features of which are contained in both these answers.

8. Along with the concept of matter associated with the materialistic answer to the basic question of all philosophy, Marx and Engels considered this concept in connection with other philosophical categories which reflect on different sides the external world as the object of scientific cognition.

(a) In Engels' 'Anti-Deuhring' it is shown that since everything in the world is formed (structurally) and originated (genetically) out of matter, the real unity of the world consists in its materiality, and this is proved by a long and wearisome development of philosophy and natural science.

(b) Furthermore, Engels has explained the interrelation between matter and motion which forms the foundation of all phenomena of the world: Motion is the mode of existence of matter. Matter without motion is just as inconceivable as motion without matter.

(c) Engels has advanced a thesis that the motion of matter as any reality in general is realized in space-time forms, and therefore the motion of matter is filling these forms. He wrote that the basic forms of all being are space and time, and being out of time is just as gross an absurdity as being out of space. In his 'Dialectics of Nature' Engels wrote that the two forms of existence of matter are naturally nothing without matter, empty concepts, abstractions. He gave quotations from Hegel's 'Naturphilosophie': 'space and time are filled with matter ... Its essence (of motion) is to be the immediate unity of space and time ...' (Engels, 1966, p. 245).

(d) Finally, everything that occurs in the world, occurs strictly within the law. As Engels emphasized, the form of universality in nature is law.

Engels absolutely denied the reduction of matter to the set of its limiting particles that were strictly identical. This reduction would mean substitution of P in place of U, characteristic of mechanism, according to which all matter consists of identical particles and all qualitative differences can be explained by quantitative changes. Engels believed that atoms, however, shouldn't be regarded as simple, or in general as the smallest known particles of matter but that atoms are compound. Rejecting the substitution of P in place of U, Engels simultaneously stressed their unity and indivisibility interpreting them as interpenetrating opposing elements. He wrote that men deal only with various really existing substances and forms of motion (as P) while matter (as U) 'is nothing but the totality of material things from which this concept is abstracted ... words like "matter" and "motion" are nothing but abbreviations in which we comprehend many different sensorially perceptible things according to their common properties. Hence matter and motion can be known in no other way than by investigation of the separate material things and forms of motion' (Engels, 1966, p. 236). Therefore, the path of cognition of matter (U) lies through cognition (P) which is in inner unity with U.

III. Science and philosophy on the boundary of the 19th and the 20th centuries

9. Between the 19th and the 20th centuries the struggle between materialism and idealism greatly intensified expanding over the sphere of science and including modern physics. Especially active was the subjective idealism (the machism) which in physics appeared as representing the 'physical' idealism. In the same area fell the conception of energetism rejecting the atomistic theory and suggesting replacement of the concept of matter by that of energy.

As the principal argument against the concept of matter the machists and their allies have advanced a thesis to the effect that this concept allegedly became 'obsolete' and inconsistent from the point of view of the latest achievements in physics and the whole of modern science. Therefore, it is said, it cannot be preserved in scientific literature. One of the theses of certain machists, in particular, was that the concept of matter is not defined rigorously, that it is based on a logical circle: matter is called what is primary with respect to consciousness while the consciousness, in its turn, is defined as what is secondary with respect to matter.

The concept of 'disappearance' of matter, of its 'reduction' to electricity, to pure energy, to pure (allegedly non-material) motion was exploited on an especially broad scale. To support this concept, the machists and in general the 'physical' idealists referred to the discovery of the electron and the radioactive phenomena, to the variability of the particle mass depending on the velocity of its motion and to other physical discoveries that led on the boundary between the 19th and the 20th centuries to the 'most recent revolution in science'. The idealistic and agnostic interpretation of these discoveries in the sense of rejecting the concept of matter and the atomic-molecular theory associated with it has brought physics together with the whole of science into a state of methodological crisis.

10. V. I. Lenin in his book 'Materialism and Empiriocriticism' (1908) studied the situation in philosophy and science; unlike the machists and the energeticists, he followed the route of further enrichment and development of the concept of matter using philosophical generalizations of the most recent physical discoveries, instead of rejecting the concept of matter. From his analysis Lenin drew the following conclusions:

(a) The philosophical concept of matter including the total object reality has the characteristic U and denotes the whole objective reality, without any limitations.

(b) Any physical concepts of the structure and properties of matter take into account only the aspect P since these are concerned not with the whole of objective reality but only with its individual aspects and manifestations.

(c)   Therefore one should not confuse as was done by the machists the philosophical concept of matter (U) with the theories of its physical structure and physical properties (P) such as the conceptions of atoms or of mechanical mass.

(d)   The philosophical concept of matter based on characteristic U can never in principle become obsolete since it expresses the invariable ability of man to reflect the outside world; on the contrary, the physical conceptions concerning the properties and structure of matter, its types grow obsolete continuously as science develops, since the special can never reflect the studied subject in its entirety.

(e)   The machist thesis of 'disappearance' of matter, of its 'reduction' to electricity is based on a logical error because it has as its prerequisite the false identification of matter as a philosophical category (U) with the theory of its physical structure consisting of atoms, and its properties, such as the mechanical mass (P).

(f)   Actually, what is observed is not the disappearance of matter (U) but the previous limit of our knowledge of its structure and properties (P) – it is not matter (U) that is being reduced to electrons (i.e. to the new and more complete P) but the atoms (to put it more precisely, their atomic shells) that previously represented the relative limit of knowledge of the corresponding U.

This was how Lenin solved the question of the concept of matter, essentially making use of the relation between U and P.

11.   For the first time during the whole history of human knowledge Lenin has given a detailed definition of the philosophical concept of matter (U). He has shown that the most general concepts in the frame of any science can be defined (revealed) only through their relation to each other, and this is by no means a logical circle. In gnoseology such are the concepts of matter and consciousness, nature and spirit, being and thought, the physical and the psychic, the object and the subject: while being most general in the sphere of gnoseology as well, they may be defined only through the analysis of their relation to each other and first of all through determination of what is the basic, primary, determining, and what is the derivate, secondary, determinable, and why. Enumerating the essential characteristics of the concept of matter, Lenin has formulated a summary of the long struggle between materialism, on one hand, and idealism on the whole, as a trend, on the other, as well as its various versions, especially subjective idealism (which was in fashion then), agnosticism, rationalism that rejected the role of sensations in the process of cognition, and the metaphysical (mechanistic) materialism rejecting the historic nature of the process of formation of knowledge, and the vulgar materialism reducing entirely the consciousness to matter.

According to Lenin's definition of the concept of matter, it is:

(a) an objective reality,
(b) existing outside and independently of human consciousness,
(c) given to us in sensations,[1]
(d) reflected, copied by our sensations, our consciousness,
(e) our consciousness is interpreted as a product of long development of matter, as a certain property of highly organized matter.

This definition of the concept of matter has a broad property U since it includes the whole objective reality without any limitations and respectively all special concepts of the physical structure of matter, its types and properties including those which are just being formed on the basis of the most recent scientific discoveries. It satisfies all the known forms of matter as well as any newly discovered ones regardless of the physical nature and the inner structure, the special properties and characteristics.

12. The failure of machists and, in general, 'physical' idealists follows, in particular, from the fact that they confuse the philosophical concept of matter taking into account U with the physical concepts of its properties and structure based on P. In his book 'Materialism and Empiriocriticism' Lenin emphasized that "it is absolutely unpardonable to confuse, as the machists do, any particular theory of the structure of matter with the epistemological category, to confuse the problem of the new properties of new aspects of matter (electrons, for example) with the old problem of the theory of knowledge, with the problem of the sources of our knowledge, the existence of objective truth, etc." (Lenin, 1962, p. 129). And further: "Materialism and idealism differ in their answers to the question of *the source* of our knowledge and of the relation of knowledge (and of the 'mental' in general) to the *physical world;* while the question of the structure of matter, of atoms and electrons, is a question that concerns only this 'physical world'" (Lenin, 1962, p. 260).

But if one should not confuse the philosophical concept of matter (U) with the physical theory of its structure (P) replacing U by P, one should not also separate them from each other, oppose one another since U is cognized only through P and P in its turn reveals its gnoseological essence in the form of U. Lenin stressed that the only correct formulation of the question would imply that with respect to any physical objects of nature one should ask whether they (i.e. the electrons) '*and so on* exist as objective realities outside the human mind or not?' (Lenin, 1962, p. 261). (Here 'and so on' denotes any objects of nature arbitrarily small and simple and possessing arbitrarily peculiar natures.) The above question is constantly answered by the scientists with 'yes', and this means the victory of materialism since the concept of matter 'epistemologically implies *nothing* but objective reality existing independently of the human mind and reflected by it' (Lenin, 1962, p. 261).

The impossibility of replacing matter as a philosophical category by the physical concept of the set of its particles logically followed in Lenin's works from the assumption that there are absolutely no limiting particles of matter, that "the electron is as *inexhaustible* as the atom, nature is infinite ···" (Lenin, 1962, p. 262). The atom and the electron are but the finite boundary marks on the infinite path of science's penetration into the depths of matter. Lenin developed these thoughts in his 'Philosophical Notebooks' (1914) and formulated the problem of applying Hegel's thesis of the unity between the finite and the infinite to the electrons emphasizing in general the infiniteness of matter deep within. The subsequent development of physics has absolutely confirmed the correctness of Lenin's forecast of the inexhaustibility of the electron.

13. The concept of matter considered in the system of philosophical categories describing various aspects of the world appears in Lenin's works as a recognition of the fact that the world is matter in motion. The world is matter moving in conformity to law: "there is nothing in

the world but matter in motion, and matter in motion cannot move otherwise than in space and time" (Lenin, 1962, p. 175). Unifying all these definitions into one, one can say that, according to Lenin, the world is matter moving in conformity to law in space and time. In agreement with this we should postulate the picture of the world of moving matter. Referring to Snyder, Lenin says that the world picture is a picture of how matter moves and of how matter 'thinks'.

The beginning of the 20th century was marked by an intense struggle between materialism and idealism concerning all these theses; this struggle ended with a victory of materialism and establishment of the concept of matter in science.

(a) The idealists and the energeticists rejected, first of all, the very concept of matter saying that matter allegedly 'disappeared', that the concept of matter has lost any meaning for philosophy and science. " 'Matter disappears', only equations are left" was Lenin's characteristic of the first cause of 'physical' idealism.

(b) Idealists and energeticists separated motion and matter interpreting motion as non-material, as pure energy. The energeticism of W. Ostwald employed this separation as its main principle. Rejecting matter, the machists and the energeticists rejected the theory of its discrete structure – the real existence of atoms and molecules.

Physical discoveries on the boundary between the 19th and the 20th centuries confirmed the validity of materialism and demonstrated the inconsistency of the machist-energeticist conception. The loss of the struggle of energeticism against the concept of matter and its structure was openly recognized by W. Ostwald who wrote: "Ich habe mich ueberzeugt, dass wir seit kurzer Zeit in den Besitz der experimentellen Nachweise fuer die diskrete oder koernige Natur der Stoffe gelangt sind ··· Die Isolierung und Zaehlung der Fasionen einerseits, ... die Ubereinstimmung der Brownshen Bewegungen mit den Forderungen der kinetischen Hypothese ··· berechtigen jetzt auch den vorsichtigen Wissenschaftler, von einem experimentellen Beweise der atomistischen Beschaffenheit der raumefuellenden Stoffe zu sprechen" (Ostwald, 1908, p. VII).

In 1909 J. Perrin compiled a table of coincidences of the results of determination of the Avogadro number by different methods, which he called 'the principle of molecular reality' or 'the proof of the reality of molecules'. In 1911 E. Rutherford discovered the atomic nucleus, and beginning in 1913 Niels Bohr developed the model of the atom. The reality of matter and the inseparability of matter and motion, as well as the reality of its structural particles were demonstrated.

(c)   The machist struggle against recognition of the objective nature of space and time, their attempts to separate them from moving matter also ended in failure. An especially important role here belonged to the special principle of the theory of relativity (Einstein, 1905).

(d)   Machism also failed completely in its attempts to reject the objective nature of the world laws as those of moving matter.

IV.   MODERN PHYSICS AND PHILOSOPHY

14. It appears as if the concept of matter in modern physics and philosophy summarizes the whole path of development that this concept has made during the course of its lengthy evolution. At the higher stage of development of science we observe the above mentioned relations between the philosophical (epistemological) concept of matter (U) that remains invariable, and the fast varying physical conceptions of its structure and properties (P). Inside the frame of physics proper its total development, especially in the 20th century, occurred along the lines of the theory of the structure and properties of matter and of the forms of its motion – from individual characteristics of opposite aspects of a physical object (P) to the analysis of their unity as U; as a result the characteristics that were previously established separately and in opposition to each other now appeared just as various aspects of a single material formation. This U in the frame of physics we shall denote by a subscript: U_{ph}.

If the establishment of the initial one-sided methods of approach towards a physical object followed from the previously domineering analytical method resulting in the real opposition separating into individual opposing aspects (P), the reconstruction of their unity and interrelation corresponds to a certain stage of theoretical synthesis based on the revealing of U as the unity of the opposites. One can summarize from this point of view the development of the concept of matter over the whole history of science and philosophy, especially during the last three quarters of the 20th century. This will be shown using as an example the three trends of development in physical and partially chemical thought:

(1) The trend that led to the establishment of the interrelation between mass (m) and energy (E) and of the principles of their conservation (see below, Section 15);

(2) The trend that led to the establishment of the common nature of two fundamental physical types of matter – the substance and the field (light) (see page 159, Section 16);

(3) The trend that led to the establishment of the unity of the electromagnetic and the non-electromagnetic characteristics of the physical nature of matter (see page 161, Section 17).

The philosophical struggle that developed around these trends in physics ended invariably in the victory of materialism and the failure of the new-machist and the new-energeticist schools.

15. Lomonosov and Lavoisier in the 18th century discovered in chemistry the law of conservation of weight or mass (m) of chemically interacting substances. In the middle of the 19th century R. Mayer and other scientists discovered in physics the law of conservation and transformation of energy (E). The two principles of conservation existed in the 19th century in parallel with each other, although Mendeleev (1871) had already formulated a hypothesis to the effect that there should be some internal relation between them. In 1905 Einstein derived from the special principle of relativity the fundamental relation unifying both principles of conservation:

(1) $$E = mc^2,$$

where c is the speed of light.

This equation (the law of the relation between mass and energy, or Einstein's law) was interpreted by the new energeticists as an imaginary transformation of mass into energy, and the energy was named the only universal foundation of the world. Accordingly, Equation (1) was given the form:

(2) $$mc^2 \rightarrow E,$$

where the arrow indicated the direction of transformations in nature. The phenomenon of mass defect (Δm) in nuclear reactions which was essentially what Mendeleev referred to explaining the fact of the existence of fractional atomic weights of elements apparently could be an experimental proof of the Equation (2):

(3) $\Delta mc^2 \to \Delta E,$

where ΔE is the energy released in the nuclear process.

Identification of mass (P) and matter (U) resulted in a false conclusion of the 'dematerialization' of matter, of its transformation into energy. In reality there is nothing of the sort, and relations (2) and (3) are incorrect. The actual equation (1) expresses the inseparability of m and E, the quantitative relation between them: each mass is associated with a strictly definite amount of energy, and vice versa. Therefore, as there could be no mass without energy, there could be no energy without mass. This thesis constitutes the true meaning of the expression (1) which presents a physical concretization of the general philosophic statement (U) of inseparability of matter and motion.

At the same time we observe here a qualitative difference between the rest mass (m_0) that is inherent in all bodies of nature and the mass of motion when an object moves with a definite velocity, in particular, with the velocity c of light characteristic of electromagnetic (optical) radiation. Similarly, as was shown by S. I. Vavilov, there is a qualitative difference between the latent energy (E_0) associated with m_0 and the active, explicit energy (E_c) associated with m_c. What was erroneously called transformation of mass into energy according to expressions (2) and (3) was in reality the process of transformation of the rest mass (m_0) into the respective amount of the mass of motion (m_c) and correspondingly the process of transformation of the latent energy (E_0) into the respective amount of the energy of motion (E_c). Therefore, we observe here complete conservation of mass and energy (while taking into account the qualitative difference between m_0 and m_c):

(4)
$$\left. \begin{array}{ll} m_0 \to m_c; & (\Delta m_0 = \Delta m_c) \\ E_0 \to E_c; & (\Delta E_0 = \Delta E_c) \end{array} \right\}$$

If in relation (4) we omit both signs of equality inside the brackets, and then in the first line omit m_c and in the second $-E_0$, relation (4) takes the form: $m_0 \to E_c$ (5) which essentially coincides with expression (2). This coincidence, however, results from a series of logically and physically inadmissible transformations of expressions (4) that reveals in greater detail the contents of Equation (1).

We observe the explicit presence of theoretical synthesis realized by expressions (1) and (4): two previously separated principles of conservation are not only unified into a single fundamental law of modern physics, but the mutual relation (the quantitative one, inaccurately called the 'equivalence') between the very conserved physical quantities m and E is exposed. If one interprets the principle of conservation of m as P_m and that of conservation of energy as P_E, the essence of Einstein's great discovery in 20th century physics can be conditionally presented as follows: $P_m + P_E \to U_{ph}$ where the arrow here and in what follows means transition from P to U_{ph} in the course of the development of physics and science in general.

16. During the whole history of physics up to the very beginning of the 20th century the substance (P_s) and light (P_l) were being studied separately, as natural formations of not only different but directly opposite character. The substance was first of all the subject of chemistry while light – that of physics (optics). The substance was interpreted as having weight, i.e. mass, discrete in its structure; light (electromagnetic field) – as weightless, mass-free, as a wave-like, continuous process. Absolutely no transition between substance and light was admissible, even as a hypothesis. As a result, substance and light were separated in every respect by a strong fence. We shall show below how in the 20th century this fence was surmounted and destroyed at three points concerning the general nature of the properties of the substance and light (mass), their structure and their ability to engage in mutual transformations.

(a) The property of mass, which up to the end of the 19th century was considered to be the monopoloy of substance, was discovered to be an attribute of light in the course of experimental measurements of the pressure of light by P. N. Lebedev (1900). But only objects having mass can exercise pressure. Later the Einstein law (1905) stipulated as a generalization that every formation characterized by energy E at the same time possesses mass $m = E/c^2$. This was accompanied by the establishment of a qualitative difference between E_0 and E_c.

Thus, the theory of light has obtained from the theory of substance the property of mass in its peculiar understanding. Following the suggestion of S. I. Vavilov, one should assume that the determining property of substance should be the presence of mass of its particles m_0 while that of light – the absence of m_0 and the presence of just m_c.

(b) At the same time, in its turn, the theory of substance exercised an influence over the theory of light (optics), in the part that was concerned with the structure of light. M. Planck (1900) has formulated the quantum theory according to which the radiation (thermal, optical) occurs not in the form of a continuous flux, but through discrete portions (in an atomic way), and introduced a quantum of action h. Later Einstein (1905) introduced the concept of the light quantum (the photon) and thus revealed the atomic nature of the structure of light (radiation). This was how the atomism that had previously held the monopoly in the theory of substance was generalized to include light. At the beginning, however, optics (up to 1923) was separated into two opposite parts: radiation and absorption of light was interpreted as occurring discretely while propagation of light – as a continuous, wave-like process. On the other hand, the substance (including the electrons, etc.) was treated as possessing an atomic structure. This contradiction was resolved through the appearance of quantum mechanics when Lois De Broglie (1923) for the first time put forth the idea that in the sphere of microscopic processes we associate with every corpuscle (particle) a certain wave and with every wave – a certain particle. This time optics has exercised the influence over the theory of substance which has acquired, along with the corpuscular structure and in the unity with it, the wave structure as well (the latter was considered to be the monopoloy of the theory of light).

(c) At the beginning of the thirties, in connection with the discovery of the positron, it was observed that hard gamma-radiation (in the field of a heavy nucleus) can produce a pair of substance particles – the electron and the positron ('pair production'), on the other hand, when the positron collides with the electron, both particles fuse and turn into photons ('pair annihilation'). This meant the disappearance of the last point of the fence previously separating substance and light. But the neo-machists and the neo-energeticists, confusing different concepts, made an attempt in this case also to demonstrate the 'disappearance' of matter, its 'transformation' into pure energy. With this purpose in mind they identified substance with matter, and light with energy. In this case the 'pair production' could be easily interpreted as the creation of matter by energy while the annihilation of particles – as annihilation (destruction) of matter. The absolute inconsistency of such attempts at reconstruction in a new manner, the Ostwald energeticism, is obvious: this whole theory is based on a logical error – the substitution of P in place of U. This is why the result of these attempts was the same as when Ostwald himself was forced to acknowledge publicly his failure.

According to S. I. Vavilov, the substance (P_s) and light (P_l) are considered as two basic kinds of matter. The theoretical synthesis in this sphere of physics can be presented as follows: $P_s + P_l = U_{ph}$.

17. Over the last four centuries in science there appeared and followed each other in succession three conceptions that interpreted the physical nature of matter and that of its transformations in a different manner. Accordingly, there were three different pictures of the world formed: the mechanical one, the electromagnetic, and modern, unified.

(a) The first, mechanical conception of matter appeared in the 17th century and existed as somewhat transformed nearly up to the end of the 19th century (approximately up to 1895). It was based on the concept of mechanical mass which later was interpreted as the rest mass (m_0). Dalton (1803) had adopted the mass of atom (the atomic weight) as the determining property of a chemical element. Mendeleev, using this property as a foundation and implying the functional dependence on it of all other properties of elements, discovered the periodic law (1869).

(b) The second, electrical or electromagnetic conception of matter had been conceived as early as the beginning of the 19th century and became established in physics in the first third of the 20th century. It was developed in the 19th century in the electrochemical theory of Berzelius, in Faraday's theory of electrochemistry and electromagnetic fields, in Maxwell's electromagnetic conception of light, in the theory of electrolytic dissociation of Arrhenius with its central concept of the ion (an electrically charged fragment of a molecule). But the decisive role here was played by Hertz' discoveries of electromagnetic waves and their application for construction of radio by Popov (1895), by the discovery of X-rays (1895), Becquerel's detection of radioactivity (1896), by the discovery of the electron by J. Thomson (1897) and others. As was shown by Rutherford, all three types of radioactive emission can be explained as consisting either of particles of substance carrying an electric charge of certain sign or as representing hard electromagnetic radiation; such an explanation was therefore totally consistent from the viewpoint of the electromagnetic conception. After Rutherford's discovery of the atomic nucleus (1911) N. Bohr created the model of the atom in the framework of the general conception of matter. If in the previous theories the determining property of matter was believed to be their mass, (m), now it was replaced by their electric charge $(\pm Z)$. When the electron was discovered, its mass turned to be variable, depending on its velocity. Therefore, it was ascribed not to mechanical but to electromagnetic origin (this is what was later called mass of motion).

(c) The third, unified conception of the physical nature of matter has appeared on the boundary between the first and the second thirds of the 20th century. Pauli's formulation of the neutrino hypothesis (1931) and especially the discovery of the neutron by Chadwick (1932) and later of other electrically neutral elementary particles destroyed the very basis of the already obsolete electromagnetic conception since it turned out to be a one-sided solution of the problem. The result of this was the return to the recognition as the more fundamental, basic property of the microscopic particles of their rest mass (m_0) which was ascribed to non-electromagnetic nature. The discovery of deuterium and the deuteron has demonstrated the dependence of the chemical properties of the elements not only on the general structure of the electron atomic shell, but also on the difference on the value of the mass of the atomic nucleus (in the case of isotopes of hydrogen and other elements). At the same time physics has preserved the positive achievements of the first third of the 20th century in the frame of electromagnetic conception, but in combination with the new unified theory taking into account along with electric properties (charge) the non-electromagnetic physical nature of matter and its properties (mass) as well as other newly discovered and quite unusual characteristics of elementary particles that could not be reduced to electromagnetism.

Thus, in this case also the progress of 20th century physics can be represented by a formula $P_m + P_e \rightarrow U_{ph}$, where P_m and P_e denote special one-sided conceptions interpreting the physical nature of matter either only in mechanical (m) or only in electromagnetic (e) sense.

18. Summarizing the above, we can say that for the analysis of the evolution in history of the concept of matter in philosophy and in science (first of all, in physics) great significance belongs to categories of universality (U) and peculiarity (P) which should never be confused or replace one another, nor become separated or opposed to one another. It is only in their unity that one finds the possibility of correct understanding both of the preceding path of the development of the concept of matter and its trends, and of the essence of the modern state of the whole problem, the essence of its formulation in philosophy and physics today. As in the general case U is cognized through P, we can say that U in its turn helps to understand and explain the essence of P which is contained exactly in this U. On the whole, the concept of matter when regarded from the point of view of total human thought developed in the direction of ascent from I to P and from P to U. In particular, the transition from separated P to its unity appearing in physics as U_{ph} is responsible for that kind of theoretical synthesis which penetrates the whole of the 20th century science and, in general, the whole of modern scientific cognition. In physics this synthesis can be expressed by a formula $P_1 + P_2 \rightarrow U_{ph}$ where P_1 and P_2 denote special different and sometimes even opposite aspects of the studied object. The essential element of this synthesis is the unified relation between the modern science (physics in particular) on one hand, and the progressive materialistic philosophy, on the other. This mutual relation reveals itself in especially complete form in the analysis of evolution of the concept of matter – one of the fundamental concepts both for the whole of science and for materialistic philosophy.

NOTE

[1] Directly or indirectly, through a device – B.K.

BIBLIOGRAPHY

Engels, Frederick: 1966, *Dialectics of Nature*, Progress Publishers, Moscow.
Kedrov, B. M.: 1970, *Dialectique, Logique, Gnoséologie: leur unité*, Progress Publishers, Moscow.
Lenin, V. I.: 1962, *Collected Works*, Vol. 14, Foreign Languages Publishing House, Moscow.
Marx, Karl and Engels, Frederick: 1973, *Selected Works in Three Volumes*, Vol. 3, Progress Publishers, Moscow.
Mendeleev, D. I.: 1905, *The Principles of Chemistry*, 2 vols.
Ostwald, W.: 1889, *Grundriss der allgemeinen Chemie*, Leipzig.

MATERIAL CAUSALITY

One of the best-remembered Aristotelian doctrines is assuredly the doctrine of four "causes" outlined in *Physics* II,3.[1] Aristotle introduces his discussion of this topic with these words:

Since knowledge is the object of inquiry, and since we do not think we know a thing until we know the "why" of it, that is, until we grasp its first cause (*protē aitia*), it is clear that we must do this in regard to coming to be, passing away, and all physical change, so that knowing the principles (*archai*) of these, we may refer our problems back to such principles.[2]

The search for "causes" is a search for the "why" of it, for explanation.[3] Thus, the distinction between the different sorts of "cause" is a distinction between different ways of answering the question "why?", i.e. between different types of explanation.[4] Furthermore, it is in why-questions in regard to *change* that Aristotle is primarily interested.[5] To specify a "cause", then, involves a reference, whether explicit or implicit, to a change. To *any* sort of change? The qualification translated above as "physical" (*physikē*) could also be read as "natural", though the examples given are mostly drawn from the realm of human making. But Aristotle is clearly thinking in terms of regularities, of things that happen always or for the most part in this way. It is not as "Polyclitus" that the maker of the statue should be described but as "the sculptor"; likewise, the person who cures, cures as a doctor; the person who builds, builds as a housebuilder not as a flute-player, and so on (*Physics* II,1–3). The physicist who seeks "causes" is looking for a pattern; he is looking for the entities that would ordinarily be responsible for an event of this sort.[6] It is not just explanation in a general or casual sense that is in question, but the kind of explanation that would satisfy the person thought to be asking the questions here, i.e. a physicist. This is not, however, to say that he is looking for a full demonstration in all cases. Though demonstration for him is through causes, the converse is *not* true: the discovery of

Butts and Hintikka (eds.), *Historical and Philosophical Dimensions of Logic, Methodology and Philosophy of Science*, 209–241.

causes need not (so far as the *Physics* is concerned) commit one to the giving of demonstrations.[7]

One other preliminary may be mentioned. It is customary to point out that the word 'cause' inadequately translates the Greek terms, '*aition*' and '*aitia*'. Since the seventeenth century, 'cause' has gradually been restricted in ordinary English usage to only one of the four types of "cause" of which Aristotle speaks, i.e. the agent (or efficient) cause. To speak of bronze as a "cause" in the making of a statue or to call natural place a "cause" of falling motion is puzzling, or even misleading, if current usage be the only arbiter.[8] But there is some reason for philosophers to challenge the restriction in explanation-styles to the one which happens to have been canonized in English usage after Hume.[9] There seems to be enough reason to maintain the tradition of using 'cause' to translate '*aitia*' in the context of philosophical discussion of explanation-styles, provided due warning is given that this older broader sense (still recognized in the OED) is the one intended.

There is a specific reason why the term 'cause' comes in useful. Though Aristotle's four "causes" correspond to four types of explanation, it does not follow that 'cause' may be translated as "explanation-type". A "cause" is an ontological factor, whereas an explanation-type is a conceptual structure. If there are four types of explanation of changes, there are four kinds of ontological factors corresponding to these types. An efficient cause, for example, is a real agency in the world, designated by the appropriate term ('sculptor' rather than 'Polyclitus'). There is no generic word in English for these factors. 'Cause' roughly corresponds to one of them, efficient cause, which would include agent-sources of all sorts, from the active point-masses of Newtonian mechanics to human agents.[10] In his translation, Hope uses the phrase, 'explanatory factor', and this is perfectly satisfactory. We shall retain 'cause': it will designate the ontological factor corresponding to any one of the characteristic modes of explanation.

This essay will be concerned with one of these, the "material" cause, which has never quite seemed to fit into the pattern suggested by the other three. The problem is to decide in what sense (or senses) the "matter" of a change can be said to *explain* the change. I propose to approach this question both from the point of view of Aristotle, and also from the point of view of contemporary science. What did

Aristotle believe to be the contribution of the "material" factor to the explanation of change? And where (if anywhere) in the patterns of contemporary scientific explanation might one expect to find a factor playing a similar role? Before engaging on these inquiries, however, it will be necessary to take a look at some of the notorious puzzles inherent in Aristotle's lavish use of the matter-concept. Since our concern in this essay is with the role of the material factor in the explanation of change, it is from his analysis of change in *Physics* I that we must begin. This is the immediate context for the discussion of the causes in *Physics* II.

1. ARISTOTLE'S MATTER-CONCEPT

The first thing to note about the Aristotle's classic account of change in *Physics* I,7 is that it is an analysis of language-about-change, rather than the empirical survey of different sorts of physical changes that one might have expected.[11] A change-situation, he notes, can be described by a pair of statements, like "man is non-musical" (*S* is non-*P*) and "man is musical" (*S* is *P*), uttered at the beginning and the end of the change, respectively. Three, and only three, linguistic categories are needed: a subject-term (*S*), like 'man', a privation-term (non-*P*), like 'non-musical', and the corresponding predicate-term (*P*), like 'musical'. These three categories are necessary and sufficient for the description of any changes capable of being specified in two propositions of this general sort.

So far the analysis is impregnable, if not especially interesting. To reject it would be to reject the possibility of describing change in the innocuous way Aristotle lays out. But awkward questions arise when we try to push the analysis just a little further. Two questions, in particular, immediately present themselves. The first, and rather obvious one, is whether this linguistic form *does* describe all changes. The second is what the *S*-term stands for: can one be more specific about the sort of ontological factor that would correspond with it?

One sort of change, and a crucially important one for Aristotle, immediately presents a problem. Substantial changes cannot be described in this fashion because there is no *S*-term which is predicable before and after. Since his analysis purports to provide the "common characteristics" of all changes, this is a serious challenge. After his discussion of "qualified" change (S comes to be P where it was non-P before) and of making (bronze-to-statue, which has some special problems, as we shall see), Aristotle gives "unqualified" change (where S simply comes to be where S was not before) only three or four lines (190b 1–5). And his embarrassment is manifest.

First, he moves from the linguistic to the observational mode: we now look at the change directly rather than at the propositional forms in which it is described, and we find (he says) "in every case something that underlies from which proceeds that which comes to be". Second, the instance he gives is that of animals and plants coming from seed, the seed being said to be the "that-which-underlies" (*hypokeimenon*). But there is all the difference in the world between the case of a man who comes to be educated and that of a plant that comes to be from seed. The seed does *not* persist in the product, whereas the man does. The plant can be simply said "to come to be" (that is what characterizes such a change as an "unqualified" one),[12] the man cannot. In the earlier analysis, something "survives through the process" as Aristotle puts it. But what survives in an unqualified change? Certainly nothing that is designated by an S-term in the straightforward way in which 'man' serves to pick out what survives the process of "qualified" change.

If something is claimed to survive in such a case, it could only be on the warrant either of an analysis of the concept of change which would discover such a continuant to be a necessary condition of change, or else of an inductive review of different sorts of change. But Aristotle does not develop either of these approaches; in the only other reference to unqualified change in *Physics* I, in a brief passage on the "underlying nature" to which we shall return in a moment, he makes use of an analogical argument to suggest that one may expect something like a material continuant in unqualified changes too. The generalization is a tentative one; we are left with the suspicion that the extension of the simple matter-form-privation triad to *all* forms of change may force a radical modification in our understanding of the functions of the different elements of the triad. One thing is clear: the linguistic mode of the original inquiry is insufficient to support a general claim about the necessary conditions of change.

Can we at least identify the S-factor in *qualified* changes? Even here, we encounter some troublesome ambiguity. What is the referent of the S-term in: "The man is uneducated"? Is it the substance, man? Or is it the "whole" man, accidents like uneducatedness and all? This question goes to the heart of Aristotle's entire theory of predication, and comes up for discussion over and over again in the *Metaphysics*. One can easily think of persuasive arguments on both sides.[13] The substance might seem to be the proper referent; accidental predication then seems to become a sort of "adding-on" of characteristics to the referent. Or one may prefer to argue that 'man' must designate everything that is in the individual entity, substance, accident, privation, in order that truth in predication be possible. We cannot follow this discussion here. But it is worth exploring the consequences of this ambiguity in *Physics* I itself.

Aristotle opens his discussion in this way:

There must always be an underlying something (*hypokeimenon*), namely that which becomes, and this, though always one numerically, in form at least is not one .... For to be man is not the same as to be un-musical. One part (man) survives, the other (unmusical) does not ... nor does the compound of the two, namely unmusical man (190a 14–21).

It might seem at first sight that the *hypokeimenon*, the substrate, that which "underlies", would be the man. The man is what persists, and the term 'underlie' suggests persistence. But this is evidently not what 'underlie' means here: the "underlying thing" is the from-which, the entity from which the change began, the unmusical man. It does not, therefore, persist. To "underlie" here means to "serve as starting-point for". The substrate is dual: it contains, in some rather complicated sense of 'contain', two principles, the continuant (the man), and the privation (unmusical). But later in the same chapter (191a 4–5), he says that something must "underlie" the contraries, i.e. a substratum that successively underlies the privation (unmusical) and the form (musical). This is to underlie in the more usual sense of this term in English, i.e. to persist. 'Substratum' (*hypokeimenon*) is evidently being used here in two different senses. The substratum-source is the composite from which the change begins; it includes the privation. It is to this composite that the change occurs, properly speaking; it is this (Aristotle says) that "becomes".[14] The substratum-continuant is the principle of continuity which makes of the event a change rather than a replacement. Specifically, it is what allows the same S-term to be properly predicated before and after change.

We shall translate *hypokeimenon* in the first sense as "subject" and in the second sense as "substratum", and refer to the context in order to decide which of the two is appropriate in a given case.[15] Aristotle notes that the subject of the change is "numerically but not formally one" (190a 15–16). To be a man is not the same as to be unmusical. The subject specifically includes the privation; this is why it cannot be said to persist through the change.

Which of these, subject or substratum, is to be identified with the "matter" (*hylē*) of the change? This is a crucial question for us, since it determines what we are to take as "material" cause. The traditional answer has always been: the substratum or continuant. Aristotle appears to be quite definite about this:

The subject of change (*hypokeimenon*) is numerically one, but is two in form. For the man, the gold, or in general the "matter" (*hylē*) is enumerable, since it is a positive something and it is not incidental (*kata sumbebekos*) that the product comes to be from it. Whereas the privation and the contrary *are* incidental to the process (190b 23–27).

The *hylē* here is the man, the gold, the positive continuant, not the complex subject (unmusical man, unshaped gold) containing the privation. The principles of change (he continues) are in one sense two (contraries), but in a more fundamental sense three, since contraries of themselves cannot act upon one another. A third principle is thus needed (as a sort of intermediary, presumably). This is the *hypokeimenon*, here understood (it would seem necessary to say) as the substratum, since the subject (which does not persist) would not explain how the contraries succeed one another, and would itself contain one of those contraries. That Aristotle intended his reasoning to be thus understood would appear from his going on to summarize as follows: "clearly there must be something which underlies (*hypokeisthai*) the contraries" (191a 4–5), which is therefore the third principle of change, in addition to the contraries themselves.

Then comes a well-known passage:

The underlying nature (*hypokeimenē physis*) is known by analogy. For as bronze is to the statue, wood to the bed, or as [matter and] the formless before it acquires a form to anything else possessing form, so is (the underlying nature) to that which has reality (*ousia*) as a being of a certain kind or an existent (191a 8–12).[16]

But here the ambiguity of '*hypokeimenon*' strikes us once more. Is the "underlying nature" the subject or the substratum here? In favor of taking it as subject would be the likening of it to the formless *before* it receives form. If this is taken temporally, then the underlying nature is the entity from which the change began (the "subject" in our sense), and not the continuant.

This is one of the arguments used by Barrington Jones in a challenging recent interpretation of *Physics* I which would deny that "matter" in Aristotle's analysis of change in this book can be taken as a persisting substratum. Jones argues that "matter" must rather be seen as the "from-which", the subject, i.e.:

that individual item, be it simply a piece of stuff or else a substantial individual, such as a seed or an embryo, with which a process of coming into existence begins, and from which the product comes to exist, where the 'from' has to be understood as having purely chronological force.[17]

There are two immediate difficulties with this reading of the passage above. First, why should analogy be involved at all in the knowing of the subject of change? Second, if the temporal sense of 'before' is asserted to be essential to the analogy, what would the prior subject be in regard to substance, i.e. to the that-which-has-reality which is the final term of the analogy? In Jones' reading it could only be *another* substance, from which this substance came to be. But this seems to miss the point of the analogy.

The root of the problem lies in Aristotle's reliance in *Physics* I on instances of human *making*. It is, of course, the kind of change we understand best, but it has a number of special features that make it a very risky analogy on which to rest a general account of change. First, craft-making does not clearly fall into either of the two main categories of change, qualified and unqualified, which he is discussing. Second, its substratum is a *stuff*, describable in terms like 'wood' or 'bronze'. But this is certainly not true of the other changes that are analyzed: when an unmusical man becomes musical, the continuant, man, is certainly not a stuff. More generally, there is no reason to suppose that the S-term in Aristotle's analysis refers to a stuff, except in the very special case of artistic shaping. But the suggestion of a "stuff" is insidiously attractive when we come to speak of the "matter" of a change.

Most important, in the case of art-making, there is no proper name for the subject from which the change begins. As Aristotle himself puts it elsewhere, the composite from which such a change originates is "obscure and nameless" (*Met.* VII, 1033a 14). The only name we have for it is the general stuff-name, e.g. 'bronze'. But this means that the *same* term is being used for the subject as for the substratum, although in different senses of course: first, as the name of an individual unshaped piece of bronze, and second, as the bronze-stuff which persists throughout the change. This equivocal usage is not possible in the case of other sorts of change, whether qualified or unqualified. It renders Aristotle's discussion of the relation of *hypokeimenon* to change irreducibly ambiguous in the context of artistic making. What "underlies" such making can either be the stuff (which persists) or the shapeless original piece (which does not), and the same term will be used for both.

Aristotle himself appears to allude to this difficulty in a much-disputed passage where he says that we ordinarily use the form '*X* comes to be from *Y*' (rather than '*X* comes to be *Y*') when we are pointing to that which does not remain; for example, the musical comes to be from the unmusical, not from man. But he then notes an exception in the case of artistic making, where one *would* say that a statue comes to be from bronze, not that bronze comes to be a statue (190a 21–26).[18] Here the bronze *does* remain, but one has no better term to designate the original unshaped piece so that the same term is used of it (even though it does not remain) as of the enduring continuant. Yet Aristotle elsewhere warns against this very usage: a statue made from gold ought to be called 'golden', not 'gold' because:

> things which come to be and pass away cannot be called by the name of the matter (*hylē*) out of which they have come to be; only the results of alteration retain the name of the substratum whose alterations they are (GC, 329a 18–20).

Even to say that the statue is made "from gold" can be misleading; though we do use this form of speech, we ought not do so without qualification, "because coming to be implies change in that from which a thing comes to be and not permanence" (*Met.* VII, 1033a 20–22), and the gold *is* in a proper sense a persisting continuant.

The merit of Jones' essay is his underlining of the fact that the *hypokeimenon* of which Aristotle speaks must frequently be understood as the subject, the from-which. But the weakness of his case is that he does not show why *hypokeimenon* and *matter* are to be equated. And even if they *were* taken to be equivalent, 'matter' would then have the same ambiguity as '*hypokeimenon*' has, i.e. it would sometimes be subject, sometimes substratum, depending on context. One could in no circumstances come to the unqualified conclusion of Jones' paper that the matter "is precisely what does not remain".

But an even stronger case can be made against Jones' thesis. Aristotle himself insists in these chapters that the subject is a composite, part of which survives – which he calls the "matter" (190b 24) – and part of which does not. The matter-form-privation solution he proposes for the puzzles about change that perplexed his predecessors rests on this distinction between the subject-from-which and the matter. The subject-form-privation triad would not do to make this point. Further, if (as Jones himself admits), there *is* a persisting element in change, like the man who is first unmusical and then musical, where could it appear in Aristotle's analysis of change except as the "matter"?

The difficulties in interpreting the matter-principle of *Physics* I are due, in sum, to two peculiarities of the text. One is the ambiguity of the term 'underlie', since both the subject-from-which and the substratum-continuant are said to "underlie" the change, and both are called by the same name, '*hypokeimenon*'. The second is the linguistic basis of the original analysis, and its consequent inability to handle unqualified change, where there is no continuously predicable name-term to designate the substratum-continuant.

It will not be necessary to dwell in the same detail on the other contexts in which matter is discussed by Aristotle; our aim was to provide a context for the analysis of matter-cause in *Physics* II. The discussion in *De Generatione et Corruptione* is the fullest and clearest. He is here criticizing the various views of matter held by his predecessors and concludes:

Our own doctrine is that although there is a matter (*hylē*) of perceptible bodies (a matter out of which the so-called elements come to be), it has no separate existence, but is always bound up with a contrariety .... We must reckon as an originative source (*archē*) and as primary the matter (*hylē*) which underlies (*hypokeisthai*), though it is inseparable from the contrary qualities. For the hot is not matter (*hylē*) for the cold, nor the cold for the hot. Rather the substratum (*hypokeimenon*) is (matter) for them both (329a 24–32).

There can be no doubt about the role of the matter-principle here: it is the persisting substratum of all change, including the unqualified changes of one element into another. He is now able to handle the problem of substantial change directly, since he has moved away from the linguistic level of *Physics* I,7. The substratum of such change is not a substance; even though the same qualitative predicate might be predicable before and after the change (*transparent* is an example he elsewhere gives, in regard to the change from air to water), it is not predicable of the substratum as such, for this would make the change into mere alteration. He concludes:

Matter (*hylē*), in the most proper sense of the term, is to be identified with the substratum (*hypokeimenon*) which is receptive of coming-to-be and passing-away. But the substratum of the other (i.e. the qualified) kinds of change is also, in a certain sense, "matter", because all these substrata are receptive of contrarieties of some kind (320a 2–5).

Matter is thus the answer to the central problem of the *De Generatione*. As substratum, it is revealed as "the material cause of the continuous occurrence of coming-to-be, because it is such as to change from contrary to contrary" (319a 18–20). One can still see the linguistic origins of the doctrine. Change is a transition from one contrary to another, and the relation of contrary to substratum is the ontological equivalent of the relation of predicate to subject. When air becomes water (i.e. when condensation occurs), there is no thought of explaining this change by means of postulated microentities, as the atomists had already suggested. Rather, it is a transition from the hot-wet to the cold-wet, where each of these properties is thought to be simply predicable, and where their ceasing or coming to be is thought to exhaust the reality of the change.

Metaphysics VIII, IX, and XII do not add much to this, except for an explicit treatment of matter as potency. *Metaphysics* VII,3 provides the well-known discussion of matter as the subject of predication. There is no reference to change. To "underlie" is to underlie predication, and predication sounds almost like an adding-on: where bronze is the matter and shape is the form, Aristotle seems to say that the "form" here is predicable of the "matter". But it is clearly of the *composite*, the statue (i.e. that which exemplifies form) that it should be predicated.[19] He even claims that substance is predicable of matter (1029a 23–24), of an "ultimate" which of itself cannot be assigned any positive characteristics. This troubling assertion has puzzled generations of commentators.[20] What sort of predication would this be? What would be an instance of it? How could such a "matter" be denoted by the subject-term of the proposition in which the predication is expressed? If this "matter", which is said to be "what remains when all else (save substance) is stripped away" (1029a 11–12), is now to be interpreted as the principle of individuation, as that which makes this object *this* spatio-temporal instance of the forms it instantiates, much more argument is needed. Aristotle shows himself properly hesitant about the line of reasoning that leads to this odd conclusion (1029a 27).

The ambiguity of the role played by "matter" in the *Metaphysics* is due to Aristotle's implicit equating of the subject of predication with the substratum of change. If change is seen as one contrary succeeding another, the substratum remaining the same, the substratum can just as easily be taken to be that which "underlies" the predication as that which "underlies" the change. And of course, this means that one can speak of "matter" (as Aristotle does in *Metaphysics* VII) even where there is no change involved. It may be argued that there is an implicit reference to change in such cases, either to the becoming whereby such a predication would come to be true in the first place, or else to the changes in terms of which such a nature discloses itself. But this is cumbrous, and in any event quite remote from the "material" explanatory factor Aristotle is seeking in the *Physics*. Let us now turn back to this, keeping in mind the multiple ambiguity of the root-concept of matter.

2. ARISTOTLE'S MATTER-CAUSE

In its first introduction, the matter-cause is defined as:

That out of which a thing comes to be, while itself remaining, for example, the bronze of a statue and the silver of a bowl, and their kinds (194b 23–26).[21]

Later, he gives some examples:

Letters are the "causes" of syllables, materials of manufactured products, fire and the other (elements) of bodies, their parts of wholes, and premises of conclusions, all in the sense of that-from-which ... "causes" therefore in the substratum sense (195a 16–20).

This is a pretty diverse list. And the emphasis on change in the first passage has become a more general reference to *origins*, to that from which the entity has come to be the kind of entity it is here described as being.

But now the ambiguity of the substratum-notion, discussed above, returns to trouble us once again. Is the material cause the that-from-which, the subject from which the change began, or is it what remains of this subject as a persisting continuant? The craftsman examples are unhelpful, for the reason we have already seen: 'bronze' serves as the name of the unshaped that-from-which as well as of the material which remains. Suppose, instead, the change is from seed to plant, or air to water? Is the material cause the seed, the air? Or is it an underlying matter, constituent of both seed (or air) and plant (or water)?

The examples he gives are not as helpful as they might have been. Letters are the constituents of syllables; they retain their identity as letters though constituting the larger unity. But fire and the other elements do not retain their identity in bodies like plants. Nor is it even true that plants come to be from the elements; they come to be from seed. The sense in which plants and other complex substances are "composed" of the four elements is not a simple one, therefore, since they neither come to be from them, nor are the elements their constituents in the sense in which flesh and blood might be said to be the constituents of animal bodies. Premises as "material cause" of a conclusion is even more confusing, since premises are not constituents of a conclusion, nor do they change into a conclusion. The from-which here is the logical from-which of inference, not the material from-which of change.

Focussing on physical change (which Aristotle announced as his original theme), there are two difficulties about specifying the concept of material cause. One is that we might try to answer a question about the "what" of a change either by determining what it was to which the change occurred, i.e. the from-which, or else by indicating the persisting substratum. The difference between the two is crucial (except in the artificial case of craft-making), because the former does not survive in the product, the latter does. To "explain" in terms of the former would be to specify origins, to point to that from which this kind of thing customarily comes. To say that plants come from seed, or that bronze comes from combining the elements in an appropriate way, tells us something important about plants or about bronze.

On the other hand, to explain in terms of the persisting substratum would be to emphasize the *continuity* of what has happened; it would be to understand the change in terms of a factor common to the from-which and the to-which. The product or outcome is explained in part by the kind of material that underlay the process. Because it was bronze (and not wood), the statue has certain traits it would not otherwise have. Because the stuff underlying seed is the sort it is, the product is plant, and not, for example, animal. Is this, however, correct? Is it not because the *seed* is the sort of seed (has the sort of seed-form) it has, that the outcome is plant? Is the reference to material cause a reference to a *kind* of material continuant, whose properties are in part, at least, responsible for the outcome? To "know the material", Aristotle says, is to know, for example, of what sort of wood a helm should be made and with what operations; this is the sort of knowledge a carpenter would have (194b 5–7). To specify the sort of wood here would be to suggest that the properties of a good helm are due in part to the choice of the right wood, that the properties of the wood (not just as from-which but primarily as continuant) help to explain the proper functioning of the helm.

This directs our attention to a second ambiguity in the specification of matter in such contexts as these, one to which Aristotle himself alludes. 'Matter', he tells us, is a relative term; what is said to be the "matter" in a given case depends on what we have decided to specify as the "form" in that case (194b 10). In *Metaphysics* VIII,4, he enlarges on this. Even if one were to say that all things come to be from the same "matter", yet there would also be a "matter proper to each, e.g. for phlegm, the sweet or the fat, and for bile, the bitter or something else" (1044a 15–19). Thus, there could be a hierarchy of "matters" for the same thing; for example, phlegm might come from the fat, the fat in turn from the sweet, and all of these ultimately from primary matter (*protē hylē*). Furthermore, in many cases the matter of a thing *must* be of a certain kind; for example, a saw cannot be made of wood, no matter what the powers of the maker.

At what point, then, should the question in regard to matter-cause end? It might seem that *all* levels should be investigated; thus we might (he says) think of the menstrual fluid as a material cause of man. But basically,

it is the *proximate* causes we must state. What is the material cause? We must not name fire or earth but the matter peculiar to the thing (1044b 1–3).

So it is the *proximate* matter (in the sense of that-from-which) that is to be singled out as matter-cause, the constituents most directly informed by the form of the being, those that are regularly involved in the coming-to-be of a being of this kind. Thus we would say, for instance, that flesh and blood (rather than fire and earth) come to be a man. What is the material cause of sleep? Aristotle answers that it is not clear what the proximate matter is: it is not just man, but man in virtue of his possession of some part such as a heart (1044b 15–17). Ought wine be called the proximate matter of vinegar or the living man the proximate matter of the corpse? It would seem so, but Aristotle feels compelled to answer that, despite appearances, they are not, because the changes in question are "accidental" ones. It is the persisting substrate of the contraries which is the source of this corruptibility, and thus ought be regarded as the "matter" of such changes (1044b 34 – 1044a 2). There is more than a hint of Platonism here: Aristotle, the naturalist, ought be prepared to regard death as a process requiring a similar sort of causal analysis to that of birth or growth. The birth-change is, of course, the acquisition of a new and higher substantial form; the death-change is not. A corpse has no single form. Because composition is attributed to the matter, not to the form of man or of wine, it is the matter-substrate which (in Aristotle's view) materially explains such passings-away, where in other cases it is the prior composite, the from-which.

The ambiguity is by now a familiar one, and it recurs again in the treatment of potency that follows in *Metaphysics* XI. The notion of matter as cause does not come up explicitly in this book. But what he has to say here about matter and potency is clearly relevant to the discussion of cause in the preceding book. Potency is defined as the "principle of change in another thing or in the thing itself regarded as other" (1046a 11); it can be either passive or active. Matter can be said to be such a principle, and the matter thought of as being of a certain *kind*: for instance, it is because something is oily that it can be burnt (1046a 25). Here, "matter" is clearly the from-which, not the substrate, since the oiliness does not persist. But if it is the from-which, once again at what point in the earlier series of changes leading to an entity of this sort ought we locate the from-which, the "matter" of this particular being?

Aristotle is at some pains to attempt to answer this troublesome[22] question:

Is earth potentially a man? No, but rather when it has already become seed, and perhaps not even then. Likewise with being healed, not everything can be healed by medical art or by luck but only what is capable of it.... In a similar way in regard to what is "potentially" a house: if nothing in the thing acted on – i.e. in the matter (*hylē*) – prevents it from becoming a house, and if there is nothing which must be added or taken away or changed, it is potentially a house, and the same is true of all other things, the source of whose becoming is external. Where, however, the source of the becoming is in the thing itself that comes to be, a thing is "potentially" all those things which it will be of itself, if nothing hinders it. The seed, for example, is not yet potentially a man, because it must be deposited in something other than itself and undergo a change (1049a 1–16).

He concludes, therefore, that in a series of becomings such as these:

a thing is *potentially* (in the full sense of that word) the thing which comes after it in the series. For example, a casket is not earth nor earthen, but wooden. For this last is potentially a casket, and this is the matter of a casket, wood in general of a casket in general, and this particular wood of this particular casket (1049a 21–24).

The proximate "matter" of a change must therefore, be carefully distinguished from the persistent "matter". Potency resides in the former, rather than the latter. (Aristotle speaks of "potency" in several other senses which have reference to co-determinability, to limitation or to capacity, rather than to change, but we will leave aside this complication here).[23] What makes the seed capable of becoming a man is the fact that it is a *seed* of a certain sort, not just that the underlying matter of the seed-to-man change is of a certain kind.[24] It is a *substance* of a certain kind, then, that serves to explain why the next substance in the transformation-series is of the kind it is: "one actuality always precedes another in time right back to the actuality of the eternal First Mover" (1050b 4–6).

On the other hand, Aristotle is still (as we have seen) tempted to make the *underlying* matter the principle responsible for changeability-as-such, and especially for composition, i.e. for those changes where forms "lost their hold". Furthermore, where there is reference to the role of *primary* matter, it can only be to matter-as-persistent, not to the from-which (since the from-which is necessarily a "this", and primary matter is never a "this", 1049a 27). But primary matter can never as such be the potency for any *particular* change; it can never serve as a principle of explanation for why a change is of the kind it is. It can only explain how an unqualified change is possible in the first place. What, then, would be the matter-cause in such a change? Insofar as it is to be sought in the proximate matter of the change, it would be air (this particular mass of air) which would among other things determine the quantity of water that would be generated. But if the question to be asked has to do with the fundamental transformability of air into another substance, the primary matter would have to be invoked as explanatory factor.

It is time now to draw together the threads in this discussion of Aristotle's multi-faceted "material cause". Solmsen remarks that "by all odds, the identification of one cause as the 'material' represents Aristotle's original contribution to the doctrine of four causes".[25] So far, it has proved difficult to pin down just what this contribution was, the ambiguity deriving from the further ambiguities we have located in the notions of substratum, of predication, of proximate matter, and of potency. It seems possible to separate at least four rather different notions of material cause in Aristotle, i.e. four different ways in which a "matter" can be cited in answer to a question about change.

The first corresponds to a rather basic question: "what was it that changed?" We recall the analysis in *Physics* I which focussed on the logical subject of predication-about-change, the S-term which is applicable throughout the change. When a man becomes musical, it is the man who has changed. Suppose, however, it is a plant which comes from seed. There is no S-term here which is applicable throughout. What is it then that changes? The seed? This would be a common answer but Aristotle has given us reason to question whether the seed can properly be said to "become" a plant. Is it the seed-stuff, which comes to be a plant after having been seed? There are problems with this answer too, as we have seen. The linguistic approach does not work with unqualified changes, or at least, let us say that it is inadequate of itself to answer the simple question about what it was that changed, the question that can so easily be answered in regard to qualified change, where the same S-term is properly predicable throughout. This is, in Aristotle's terms, a "logical" rather than a "physical" issue, and the material cause is here the substance that endures, the matter-form composite.

But is it properly called a material *cause*? It is clearly the "matter" of the change, in one of Aristotle's senses of that term. But does it *explain* the change, help to make it intelligible? It is one of the factors constituting the change-situation. But to specify it is to do no more than *describe* the change, it would seem. In contemporary usage, to "explain" is to go beyond the given, to embed the puzzling fact in a broader context of some sort. To say what it is that has changed does not offer to go beyond what one knew to begin with, if one knew that a change had occurred at all. But 'explain' for Aristotle did not, it can be argued, have this same overtone. To "explain" might well be to point to the obvious, and to mark it out as a necessary condition. Further, in this case, it is to point to *substance*, to the thing as bearer of predicates. It is to recall a theory of predication in which change comes to be described in terms of successive "contraries" (contradictories really) qualifying the same substrate. This is to go beyond the given, at least in the sense of suggesting a metaphysical structure to which one has access equally through analyses of predication and of change.

The main shortcoming of this notion of material cause is its limitation to qualified change. In the case of unqualified (substantial) change, it does not apply. One cannot say that it is the primary matter which changes; in fact, the primary matter is precisely what is seen as the principle of continuity. It is what does *not* change. In qualified change, the same answer can be given to the two questions: "what is it that changed?" and "what is it that remained the same?" The answer in both cases is "the man", when the change is a man's learning a musical skill. Aristotle's explanation of this puzzling identity of answer to what seem like contradictory questions is that it is the man that changes inasmuch as he is the bearer of successive contraries, whereas he remains the same in his substantial identity as man. But clearly this analysis does not apply when the substance *itself* changes.

The second sense of "material cause", i.e. the second sense in which matter can be said to "explain" change within Aristotle's system is as the "from-which". To specify the starting-point of the change is implicitly to assert that within the original entity there was a capacity for the change that actually did take place. Here we shift from the context of predication to that of physical potency. To say of a particular seed that it has the potency to become an oak (and not, for example, an elm) is to say something informative and important. It is, besides, to "explain" what subsequently happens when an oak *does* come from this seed. It is an *explanation*: it goes beyond the mere specification of the thing as seed. It imputes an ontological structure to the seed that a statement of its present actuality would not of itself suggest. This is even more clearly true if one takes non-organic changes such as those from element (e.g. air) to element (e.g. water), or from craft-material (bricks and timber) to product (home). One might know a great deal about bricks and timber without suspecting (if one had never seen a house made of them) that *this* was one of the many things one could construct with their aid.

The sense in which to specify the from-which is to *explain*, then, is to go beyond the given to present the from-which not as actuality but as potency for a particular type of change. In the organic realm especially, the realm in which Aristotle was most at home as a scientist, this sort of "explanation", the recognition of a web of interrelated potencies, is crucially important. There is a symmetry between material cause and final cause in this reading, for one is the starting-point of a change, the other its end. The change itself is seen as instance of a natural regularity: such a change *regularly* comes to be from starting-points of this M-sort, and *regularly* tends to terminate in an entity of this F-sort. No separate *formal* cause needs be specified; the formal aspects of the change are entirely specified by specifying the material and final causes. The material cause is viewed as an actuality of a particular sort, proximate "matter" for, or in potency to, a particular process ending in an entity of a different sort.

The third sense of "material cause" is the hardest to specify. It is matter-as-substratum, the persisting continuant, that which ensures the continuity of the change itself and makes of it change rather than replacement. This is the sense which is implicit, as we have seen, in Aristotle's stuff-analogies of bronze and wood. It is also implicit in his account of primary matter as the minimal continuant which underlies substantial change. It is that out of which the final product is constituted by imposing a form of some kind on it. The stress here is on matter as privation or on passive potency rather than on matter as *active* potency; it is on the bronze as lacking the form of the statue rather than on the original piece of bronze as capable of coming to be a statue of a particular sort. Responsibility for the formal aspect of the entity in which the change terminates is here attributed to the maker rather than to the potency of the original matter. The difference in emphasis is in large part due to the differences between the explanation-types appropriate to biological process and to craft-making. Bronze is "material cause" of the statue not because it explains the *kind* of statue it is but because it is the "makings", the given, with which the craftsman had to cope.

Several puzzles remain however. Is it not the case that the *kind* of bronze, the grain of the particular piece of wood, makes a difference to the craftsman? The form of the product will depend on the *kind* of material with which the craftsman was working (bronze rather than clay) or the peculiarities of the particular instances of this material available to him (e.g. flaws in the gem). How is one to separate the "material" and the "formal" factors here, if 'formal' is taken in its ordinary sense as that which gives structure (form, actuality) to an entity? The form of the product is in part due to the form of the material used, and a specification of this material sufficient to explain the outcome must include the relevant formal aspects. The "material cause" in such cases is not, therefore, to be thought of as over-against-form or even as over-against-the-final-form, but only as material-for-that-final-form, taking 'material' here to signify the fullest intelligible reality of the substrate in question.[26]

But, of course, when we come to substantial change, the analogies of craft-change fail right away. The suggestion that primary matter is a "stuff" is, as we have seen, a very risky one, for a stuff is identified by means of a particular stuff-characteristic, and this characteristic makes a difference to the sort of end-product one may expect when a change occurs involving the stuff. G. E. M. Anscombe writes that primary matter for Aristotle is:

the substantially uncharacterisable stuff of a change from being one to being another element. It is not in itself intelligible but has to be understood as what is capable of this change; for there is nothing to it but the capacity of being now of this, now of that substantial kind.[27]

The problem here is to know why such a principle should be called a "stuff". Is there a certain quantity of it available in a given change? This is plausible (the amount of water produced from a given body of air would surely depend on the quantity of air involved), and later Aristotelian philosophers would get into enormously complicated questions about how to relate quantity with primary matter.[28] Yet Aristotle appears to claim that in its own right no characteristics may be predicated of it, not even quantity. It is difficult, therefore, to discover any justification for supposing that it may be described as a "stuff". Insofar as it can be called on as material *cause*, it is as a necessary condition for the ontological continuity of substantial changes. It is *not* required in this role in the case of accidental changes; here substance itself is the guarantor of continuity.

One further query may be raised about the functioning of a material continuant as an answer to such questions as: "what is it made of?" or "what reason is there to suppose that it was a change and not just a replacement of one substance by another?" *Does* the material factor always remain the same? In living bodies, for instance, the materials are constantly being replaced. D. C. Williams has discussed the difficulty inherent in the supposition that it is the "material" factor that gives continuity to living change.[29] The craft analogy fails once again. There must, of course, be some sort of space-time continuity between the different stages in biological process, and this continuity cannot be provided by the form since the form does not (in Aristotle's view) individuate. But the continuity is evidently not that of a stuff which takes on different forms. Could it be the continuity of a kind of material? What is absorbed from food into the substance of the organism must be *like* that which it replaces, it could be argued. If water was a constituent of the original material factor, it will remain a constituent, even though not the same individual water-particles. If different sorts of material were to be absorbed into the organism (say that particles of limestone began to replace the living cells), a substantial change (petrification) would occur. However, this is not really satisfactory because the kinds of material that can be absorbed along the way *do* vary from instance to instance (or from stage to stage) of a particular kind of biological process.

To account for persistence through change in the Aristotelian perspective is far more complicated, then, than is suggested by the craft-analogy on which Aristotle's analysis so heavily depends. Both form and matter will be involved, and one will have to introduce (as the later scholastics did) notions of subsidiary or of virtual forms, a topic which Aristotle left in some obscurity. It simply will not do to assume that there is something of the original from-which left in the product, and that this something automatically qualifies as the material cause, in our third sense of this term. For in living process, there may well be nothing of the original subject left, nothing, that is, of the individual entities of which the original composite was composed. What has remained is the "principal" form directing the process and somehow individuated to this particular instance.

Besides the three senses of "material cause" analyzed above, there is one further sense that is quite commonly presented as *the* sense of the term in Aristotle's work. This is matter as the "constituents" of a thing, as that which "the thing to be explained is made of or can be cut up into".[30] Here is the "material" as the determinable, and the form as the determining, in an already-constituted composite. But material *cause* was postulated by Aristotle as an answer to a question about change. Insofar as the constituents of a thing could be described as its material "cause", it would seem, then, that it could only be because they had also constituted the original subject or the persisting substrate of the change by which this entity had come to be what it is. They would qualify as "cause", then, because they serve to explain this change, and not because they presently help to constitute the entity.[31] Were a living body to be cut up, the resulting parts, no longer flesh and blood, could be regarded as the material cause of a corpse (1045a 1–2), but certainly not of the original living animal. It is the craft-analogy, as usual, that misleads one into supposing that the present constituents of a body can, as such, be identified with the materials from which this body came to be, and thus with the material cause of this entity.

There is, however, one context where Aristotle undoubtedly *does* take constituents as material cause independently of any question of coming-to-be. This is in the *Posterior Analytics*. Here the matter serves as "cause" not in relation to change, but (as we have seen) in relation to demonstration of properties. The curious initial definition of it ("an antecedent which necessitates a consequent", 94a 22) which does not seem to mark it off from the other causes, is filled out by several examples of what are proposed as demonstration by material cause. In each instance, 'material cause' is the name given to the *parts* of the entity, some property of which is being demonstrated. The first is a geometrical example where his schema works very well, as might be expected, since his model of demonstration was inspired by geometry in the first place, and since the "material" is known in advance by the mathematician's own construction. A second example assumes as an hypothesis that light consists of small particles, and goes on to draw deductions from this.

A third example is worth quoting in full:

As the common properties of horned animals we collect the possession of a third stomach and only one row of teeth. Then since it is clear in virtue of what character they possess these attributes – namely their horned character – the next question is ... (98a 17–19).

This connection between lack of upper teeth and the possession of horns he endeavors to show in another passage:

Thus it is that no animal that has horns has also front teeth in both jaws; the upper teeth are missing. For nature, by subtracting from the teeth adds to the horns; the nutriment which in most animals goes to the former being here spent on the growth of the latter (*De Partibus Animalium*, 663b 36–664a 3).

The reasoning is no more demonstrative than the instances he cites of the three other "causes" serving as middle term. But what he has in mind is clear: from a knowledge of the functioning of the constituents of a thing, one may be able to infer some properties of the thing itself taken as a unitary whole. This is to explain a feature of the whole in terms of the parts. This gives us an acceptable notion of material cause, which though it prescinds from the context of coming-to-be, the context in which the problem of explanation was originally posed, is more directly linked perhaps to the techniques of scientific explanation than are the other three types of material cause we have investigated.

Before leaving this tangled question, there are two further notions of material cause that have been proposed by readers of the Aristotelian text. The first is matter as the principle of individuation, "explaining" how a universal may be endlessly instantiated. This is a Platonic, rather than an Aristotelian, theme; indeed, Aristotle hardly ever alludes to it directly. Later Aristotelians developed an account of individuation from the discussion of substance in the *Metaphysics*. Plato maintained that form needs a matrix to allow it enter the arena of image and becoming. The analogy he uses is closer to that of a space than of a stuff. None of Aristotle's various matter-concepts seems to correspond to this one. And the functions that Aristotle attributes to a matter-principle (or more correctly, to a group of principles loosely called by the same name) are not, any of them, easily reduced to that of individuation. Both the subject-from-which and the underlying substance might be said to individuate. To make primary matter *the* principle of individuation (as indeed the later Aristotelian tradition tended to do) is not so simple, and certainly requires a reconstructing of the original argument to primary matter in the context of substantial change. But to discuss this would lead far afield.[31] And in any event, it seems fairly clear that this was *not* one of the senses in which Aristotle was postulating a "matter" factor as one mode of explaining change.

There are some passages in which Aristotle proposes the genus as "matter" of the thing defined (e.g. 1038a 5–9). Thus, animal would be "matter" of man; the relation of genus and differentia is therefore proposed as analogous to that of matter and form. There are two ways of taking these passages. One is to suppose that the use of the term 'matter' here simply suggests that the differentia determines (or limits) the genus somewhat as form is taken to limit matter. Though this makes no reference to coming-to-be, it might perhaps be taken to define a material "cause" in all scientific definition. One would, therefore, be alluding to an invariable feature (genus) in Aristotle's account of explanation.

Rorty, however, argues for a bolder claim, namely that Aristotle means the genus to be taken as the proximate material cause of an entity, a doctrine with which Rorty himself disagrees. Only in this way (he suggests) can one arrive at a plausible overall interpretation of *Metaphysics* VII–IX, and specifically of the relation between definition and form hinted at in such puzzling passages as: "The proximate matter and the form are one and the same thing, the one potentially

and the other actually" (1045b 17–19). He concludes that in response
to the question of whether to take the whole definition or only part of
it to correspond with the specific form of a thing:

Aristotle has to say something which he never says explicitly but which he must maintain
if he is to regard H(VIII), 6, as his answer to the question of Z(VII), 12. He has to say
that genus, in a definition by genus and differentia, somehow signifies the material cause
of the individuals of the species defined. In other words, he has to say that e.g. animality
stands to rationality as the brass of a statue stands to its shape. The proximate material
cause of a man is that sort of organic material which can be called "animal" but cannot
be called "human" .... The genus is a name for the sort of thing that an exemplar of a
species of the genus can be made out of.[34]

The difficulties with this interpretation are considerable. To take the
genus-term to be the name of a sort of stuff ("undifferentiated animal
goo")[35] which is in potency to the specific form, and which will, upon
being imposed, convert the "stuff" into an instance of the species in
question, seems to run contrary to much of what Aristotle has to say
about the proximate matter of a change. This matter is ordinarily
described (as we have seen) as a substance in its own right, a seed, for
instance. It is not just undifferentiated plant or animal material, but a
seed (which is, of course, not the genus of the grown organism). To ask
about "material cause" in the sense of the proximate matter of the
change by which a being of this sort would come to be, tends to lead us
to consider only those processes by which kinds of thing regularly
come to be. For Aristotle, these are primarily living processes and
craft-making. The material of the latter might, it is true, be taken to be
a sort of "genus" (bronze, for example, as the genus of the bronze
statue).

In the case of living processes, however, this will not do. If Aristotle
had believed that organic genesis begins from a "goo" for which the
genus-term of the adult organism is one appropriate name, he would
surely have said so. To portray organic growth as a progress from an
undifferentiated generic starting-point (plant-in-general) to a progres-
sively more specific outcome (oak tree) does not do justice to the texts
where Aristotle speaks of it as a progress from a matter which has a
specific capacity to develop into the kind of being that *does* in fact
appear. If the matter were the undifferentiated generic nature only
(animality), the explanation of the process to the *specific* outcome
(man) would have to rely on the efficient cause only. It would be as

though from the same material cause *any* species within that genus could, in principle, develop. This seems so remote from anything that Aristotle, the biologist, might have held, that stronger grounds for it than Rorty is able to adduce would seem to be called for.

We shall, therefore, limit the senses of 'material cause' in Aristotle to four only. And now to supplement this historical review, it would be worthwhile to inquire very briefly as to whether any analogues of these four may be found among the patterns of explanation in the natural sciences today.

3. MATERIAL CAUSALITY IN CONTEMPORARY NATURAL SCIENCE

First, we may rule out any factor which can be arrived at by an analysis of the description in ordinary language of a change. Such a description would not be accepted as explanatory; it could at most only serve to specify what it is that is to be explained. And the logical technique of asking which terms are predicable throughout would scarcely be acceptable as a way of revealing a physical substrate.

A second approach would be to postulate an indeterminate physical substratum of all change which would be analogous to primary matter. One thinks of the ether of Maxwell's theory, "the subject of the verb, 'to undulate'" in Eddington's memorable phrase, or of the energy that Ostwald supposed to underlie all physical change and structure. When the contemporary physicist specifies the electric field-strengths at a point in space, he does so in terms of certain hypothetical operations and the measure-numbers they would yield. If someone were to ask him: but of what are these numbers *predicated*? what is it that is *there*?, being more cautious than his nineteenth-century counterpart, he would answer: I do not know, and I am not sure that the question is even meaningful. By hypothesis, it cannot be answered in terms defined within the theory, for if it could, the "matter" would no longer be an indefinite substratum. And so it cannot be talked about in a meaningful way within the theory.

The tendency today is, therefore, to refuse the question: "but what is it made of?" when applied to fundamental physical entities. One can say that a molecule is "made up of" atoms, or an atom of electrons, protons, and the like. But ultimately there is a limit to this sort of question. For the moment, at least, it makes no sense, for instance, to ask: of what sort of "stuff" are electrons made? or what *is* electrical charge? To this question, a "matter"-answer is required. One hardly ever hears this sort of question today, though a century ago it was constantly being put. This seems to signify a gradual rejection of the assumption that matter-stuff analogies could furnish the ultimate in physical explanation.

Ought we, therefore, retain a "material" factor of this sort in our theory of explanation? It would play no specific role in particular explanations, no more than primary matter did for Aristotle. It would serve only as a "horizon" reminder, a way of underlining that physics does not reduce to mathematics, that beyond every mathematical structure postulated by the physicist there is the inexhaustible physical referent and the surprises it may still have in store. 'Matter' here refers not to a featureless substrate or to a principle of continuity, but to that which lies beyond the present level of explanation. To say of physical explanation that it has a "material" aspect would be shorthand for saying that it is always provisional, always open to further sharpening.

When we turn to the use of 'material cause' to denote the potency of the original subject of a change, we are reminded of the central role played by genetic explanation in contemporary science. It can take two forms, developmental (ontogenetic) and evolutionary (phylogenetic). The former of these Aristotle would have found congenial, the latter, not. In a developmental explanation in embryology, for instance, the manner in which an individual of a particular kind comes to be is traced. The "material cause" here is the entity from which the process begins, considered as the potential source of that process and of its outcome. Thus the DNA of a fertilized egg, for instance, is described as the "blueprint" of what is to come after. The metaphors of code, information, and so on, are the contemporary way of signifying what Aristotle meant by potency in this context. Biologists as yet have only the faintest clues as to how the protein synthesis initiated by DNA can give rise to the immensely complex progressive differentiation of cells outward from the original single cell. But they *can* say that the basic information is contained in that cell.[36] This is already a form of explanation, even though the developmental pathways are still almost entirely unknown. It is extremely important to know in what elements the potencies for the process to come actually lie, and it is often a matter of delicate scientific analysis to determine, for instance, the relative roles in this regard of the physiological entity and of its environment.

Evolutionary explanation is of a much more basic sort. It answers the question: how did this *kind* of being come to be in the first place. Aristotle, who made the permanence of species the basis of his entire theory of cosmic order, rejected this mode of explanation when it was proposed by Empedocles. The "material" factor in such explanations can be taken to be the organism to which some segment of the evolutionary process can be traced. This organism is once again considered precisely as the bearer of the "potency" for the subsequent process. On a broader level, the "material cause" might be taken to be an entire context, organic and inorganic, to which the process can be traced.

There is a crucial difference between the two types of explanation: one is predictive and the other is not. From a knowledge of the material factor in a developmental situation, one can in principle predict the outcome,[37] though in practice this is as yet not possible for even the simplest organism. From a knowledge of the "material" factor of an evolutionary sequence, one cannot, however, even in principle predict what will happen. This limitation on the possibility of evolutionary prediction does not imply that evolutionary *explanation* cannot be given. (It was this asymmetry between prediction and explanation in the context of evolution that constituted from the beginning one of the most serious objections to Hempel's "symmetry" thesis). The biologist works backward from the end-product of the sequence, and tries to discover plausible steps (sufficient conditions) by which this end-product could have been developed. In the developmental case, the "material" factor can be discovered even before the pathways leading from it to the adult organism are known. But in the evolutionary case, it is only by tentatively tracing the pathways backwards that the "material" factor can be discovered. The specification of the "material" factor (i.e. of a stage in the changes leading to the organism under investigation) can never be better than tentative.

Much more could be said about the "material" factors in both of these types of genetic explanation, but perhaps enough has been said to indicate that an analysis of the structures of explanation in biology would reveal not just one, but two, aspects of these structures which bear some resemblance, at least, to the material cause of long ago.

Finally, and perhaps most significantly one might mean by 'material cause' the "given" elements in a structural or dynamical explanation, the "materials", if you will, that the theoretician has at his disposal. These will ordinarily be hypothetical entities whose behavior is postulated. The material cause in the kinetic theory of gases would thus be molecules, in Newton's mechanics, mass-points, in early atomic theory, protons and electrons, and so on. At any given stage of science, the "material" elements are those about which no further question is as yet being entertained. Kinetic theory could not have answered a question about the internal structure of gas-molecules, nor can fundamental particle theory as yet be sure that it can go beyond protons, electrons, mesons, to quarks of color, charm, and strangeness.

Aristotle's craft-analogy now begins to seem more helpful. A natural change is not the product of art, but the *explanation* of a natural change is. The scientist chooses certain "materials" with which to construct his model. Like the craftsman, he chooses the materials that will give him the model he needs. Once he has chosen the materials, he keeps on working with them, unless and until it appears that he cannot with their aid construct what he requires. The "material cause" here is the set of entities of which the model is constituted. The construction of a theory and its associated model will require one to specify a set of hypothetical entities and the laws governing them, and then to relate these entities in a structure of some sort (formal cause).

Not all explanations in the natural sciences fall precisely into this pattern. But it is fairly pervasive. The "material" aspect of it was badly misunderstood in the early years of the Scientific Revolution; it was supposed that the matter of the explanations being offered in the new mechanics could be exactly specified in terms of non-problematic primary qualities. Only the hypothetical structures in which the materials were linked, and the derivation from them of secondary qualities, seemed problematic. It was not, indeed, until our own century that the problematic status of the "matter" of scientific explanation, i.e. of the entities taken as "given" at a particular stage of theory, became evident. Much of the discussion of conceptual change in science which has gone on among philosophers in the last decade bears on this issue, but perhaps enough has been said to make it unnecessary to develop this in any more detail.

Aristotle began his analysis of the explanation of change from a consideration of the craftsman and his bronze. The wheel is now full circle: we are back with craftsmen shaping their materials to make the intricate structures of physical explanation itself. The great difference is that the bronze of Aristotle's artist really *was* a "given"; there was not much he could do to alter its nature. Whereas the "matter" of physical explanation today is a creation of the physicist's own imagination, not a free creation to be sure, but one where the physicist always retains something of his rights as creator.

University of Notre Dame

· NOTES

[1] An earlier version of this paper was presented under the title: 'Whatever Happened to Material Causality?' at the annual meeting of the Metaphysical Society of America in Atlanta in 1963.

[2] 194b 17–23. The translations throughout are my own. In making them, I have consulted the standard Hardie–Gaye translation of the *Physics* as well as the more recent ones by R. Hope and by W. Charlton. The reader should be warned that there are wide – and substantive – differences between these translations in regard to some of the passages discussed here.

[3] The word 'why' in English is rather more restrictive perhaps than the Greek '*dia ti*'; Charlton translates it as "on account of what?", which is suitably unspecific.

[4] See *The Concept of Matter*, ed. E. McMullin, Notre Dame, 1963, pp. 17, 192; J. Moravcsik, 'Aristotle on Adequate Explanations', *Synthese* **28** (1974), 3–17.

[5] He does allow that a "why?" may be put to "things which do not involve motion as in mathematics" (198a 17). But this is clearly the exception; his analysis ordinarily assumes that "causes" are "causes" of change (including coming-to-be).

[6] One class of events, he recognizes, do not readily fall into the category of "the necessary and the normal" (196b 20), yet they may be said to be "for the sake of something" and thus to lend themselves to causal explanation (rather than being relegated to the category of chance). These are events that depend on deliberate human decision. To explain in terms of reasons does not commit one to regularities nor to well-defined potencies like house-building. Aristotle does not pursue this theme unfortunately; it would quickly have led him to a terrain of reason/cause that has recently come in for much discussion. For the most part, he takes it that he is analyzing natural processes where "the sequence is invariable, if there is no impediment" (199b 25–26).

[7] In a recent article, Max Hocutt claimed the converse to hold: "For Aristotle, specifying 'causes' is constructing demonstrations" ('Aristotle's Four Becauses', *Philosophy*, **49** (1974), 385–399). But the material cause described in *Posterior Analytics* II,1 is quite different from the one we find in the *Physics*. And the account in *Physics* II explicitly refers back to *Physics* I. More seriously, the notion of demonstrative syllogism quite clearly does not apply to the examples of statue-making and house-building that are the norm in the *Physics*. There is nothing new about stressing the importance – and the problematic character – of the four causes in Aristotle's account of demonstration. This was a central point in neo-scholastic natural philosophy; see, for example, M. Glutz, *The Manner of Demonstration in Natural Philosophy*, River Forest (Ill.), 1956. What must be rejected is Hocutt's assertion that the analysis of causes in *Physics* II is to be understood in terms of the theory of demonstration in P.A. It is by no means "incredible" (his complaint) that the P.A. is "rarely cited" by those attempting to understand *Physics* II.

[8] Hocutt (*op. cit.*) and G. Vlastos ("Reasons and causes in the *Phaedo*", *Philosophical Review* **78** (1969), 291–325) get quite exercised about the impropriety of this usage. Yet it can be argued that philosophers, like Aristotle, quite frequently appropriated terms in ordinary use for special functions in their philosophy, stipulating the requisite senses as they went. There is no equivocation here unless the user (or the reader) forgets what has been done. It is "absurd" (Vlastos, p. 294) to speak of health as the "cause" of taking a walk only if we commit ourselves in advance to the identification of "cause" with "productive cause".

[9] John Passmore, 'Explanation in everyday life, in science and in history', *History and Theory* **2** (1962), 105–123; see p. 109.

[10] It may be noted that the terms 'mechanical' or 'mechanistic,' often used in this context (see, for example, Hocutt, *op. cit.*, p. 391) are dangerously ambiguous. 'Mechanical' might refer to the explanation-styles of Cartesian physics (which would admit only impact action as "mechanical"), of Newtonian physics (allowing in addition attraction at a distance), of quantum mechanics (permitting indeterministic modes of action), or even to some future "mechanics", suitably expanded and reformulated to allow the incorporation of the modes of interaction peculiar to living systems.

[11] Aristotle's choice of terms throughout *Physics* I underlines this linguistic emphasis: it is predication-about-change that is his concern, and the warrant for the claims he will make. See Section 3, McMullin, 'Matter as a Principle', *The Concept of Matter*; Section 2, B. Jones, 'Aristotle on Matter', *Philosophical Review* **83** (1974), 474–500.

[12] Whether the seed-to-plant process constitutes substantial change or merely growth is another question, and one incapable of decision on the criteria of *Physics* I.

[13] See, for example, the essays by Sellars, Fisk, McMullin, in *The Concept of Matter*; also P. Geach, *Reference and Generality*, Ithaca, 1962.

[14] 190a 15. It could be argued that it is not the substratum-source but the substratum-continuant that is properly said to "become". When S becomes P, it is still S. A man becomes musical, from having been unmusical. Ought we say that an unmusical man becomes musical? Aristotle thinks so, since only a specifically unmusical man *can* become musical in his view, which makes this the proper way to describe the change. Usage does not decide the issue in this case. But in unqualified change, it seems less appropriate to say: "the seed becomes a plant". The plant comes to be, and it comes to be *from* seed. But to say that the seed becomes a plant wrongly assimilates this change to a "qualified" one (like man becoming musical), and suggests that the seed persists in a new plant-like form. The ambiguity, both in Greek and in English, of the "X becomes

Y" usage (does it or does it not entail that S itself persists?) underlines the dangers of basing an analysis of change on change-language, as Aristotle does in *Physics* I.

[15] The unfortunate consequences of the ambiguity of the term, '*hypokeimenon*', in this chapter are easy to illustrate not only in commentaries on the chapter, but even in the standard translations. For instance, Hope takes 190b 20 to say that "every product is composed 'of' a persistent being (*hypokeimenon*) and a form", which sounds like a plausible statement of matter-form composition. But what the text says is that the product "comes to be from" a *hypokeimenon* and a form, and this *could* just as easily mean that the product comes to be from the subject (the original pre-change entity) and the (supervening) form.

[16] The specific reference to matter (*hylē*) is conjectural in the Greek text.

[17] *Op. cit.*, p. 476.

[18] Jones constructs an elaborate argument around his reading of this last clause. Since there is no reason in his view why one should not say that the bronze comes to be a statue (we shall see one below at 1033a 20–22), he reads instead: "we do not say that the bronze statue comes to be"; rather we would say, he takes Aristotle to imply, that the "bronzen" (made-from-bronze) statue comes to be, in order to avoid calling the product by the same name as that from which it came to be. This reading is defensible, but from it in no way follows the conclusion he draws: "The statue is made from bronze, and bronze remains, but bronze cannot be the underlying thing. For that must be the unshaped piece with which the sculptor began. Therefore, the fact that Aristotle says that the bronze remains is not sufficient to show that the matter remains or that there is any particular thing which remains. If anything, we have seen, matter is precisely what does not remain" (*op. cit.*, pp. 487–8). The subject of course does *not* remain. The bronze which remains is not a "particular thing" in the sense in which the man (who becomes musical) is; this is one of the pecularities, already noted, of stuff-changes. But why should this imply that it cannot be the *matter* of the change? Jones asserts that for Aristotle, matter is necessarily a discriminable individual, and hence cannot be equated with the bronze-substratum of the change which is (he supposes) *not* a discriminable individual. But is there not a sense in which this substratum *is* a discriminable individual, by comparison with other pieces of bronze, for instance? There seems to be no reason why it could not serve as the matter of the change, unless one simply begs the question by assuming that matter *must* mean the discriminable individual which is the subject-from-which of the change. It does not seem necessary, therefore, to have recourse to the suggestion made by Alan Code in his detailed critique of Jones' paper ('The Persistence of Aristotelian Matter', *Philosophical Studies* **29** (1976), 357–367) that the "we do not say" of the Aristotelian text above ought be replaced by "we do not *only* say." The context would lead one to *expect* Aristotle to say this, but since he does not, it must be regarded as a risky reading.

[19] R. Rorty, 'Genus as Matter', in *Exegesis and Argument* (ed. by E. N. Lee *et al.*), Assen, 1973, pp. 393–420; see p. 401.

[20] One recent attempt to make sense of it involves assimilating the ultimate subject of predication to the primary matter of substantial changes, and treating this matter as a sort of "low-grade substance". See J. Kung, 'Can Substance Be Predicated of Matter?', APA Western Division, 1976.

[21] The phrase '*ti enuparchontos*' appears to refer to a constituent which itself remains.

[22] Rorty suggests that this would be a slippery slope for Aristotle because "it suggests that instead of having the world blocked out into species of substance which can be

matter for one another, we get a potential X only after a whole series of accidental changes have brought a Y just to the brink of toppling over into being an X. So we cannot say that primary substances which are in species Y are the proximate material causes for the coming-to-be of primary substances which are in species X, but rather those Y's that have been worked up a lot ... (and this would be) well on the way to a law-event rather than a thing-nature framework" (op. cit., p. 415).

[23] See McMullin, 'Four Senses of Potency', in The Concept of Matter, op. cit., pp. 295–315. See esp. p. 307 where the common mistake of identifying Aristotelian matter (in the sense of underlying matter) with potency in the context of change is discussed at some length.

[24] Aristotle frequently identifies potency with the capacity for doing something: "all potencies are either innate, like the senses, or come by practice, like the ability to play the flute, or by learning, like artistic skill" (1047b 1–3). This potency resides in the matter-substrate, since it continues whether or not the capacity is being exercised. Nonetheless, it is as it is found in the subject prior to the change that it explains what happens. The fact that a flute-player can still be said to have the capacity to play the flute while actually playing it is not important to the explanation of the outcome; what is important is that he knew how to play it beforehand. This is what serves to explain what happened. It is the fact that a man with his eyes shut can be said to have the capacity to see that explains what happens when he opens his eyes (1048b 2).

[25] Aristotle's System of the Physical World, Cornell, 1960, p. 123.

[26] G. E. M. Anscombe argues that the "material aspect" of changes such as bronze-to-statue cannot be spoken of, because to speak of them would of necessity introduce the formal aspect (Three Philosophers, London, 1961, p. 53). But Aristotle never does say this, and in fact makes rather a point, as we have seen, that saws for example have to be made of a certain sort of "material" if they are to function as saws. To specify the "material aspect" of a stuff such as bronze is to specify it as bronze. The passage quoted as "The form can be spoken of, and anything can be spoken of qua having a form, but the material aspect as such, never" (1035a 7–9) occurs in a chapter where the question at issue is whether the matter and form can both be called substances. Ross translates the passage as follows: "For the form, or the thing as having form, should be said to be the thing, but the material element by itself should never be said to be so".

[27] Op. cit., pp. 52–3.

[28] J. Bobik, 'Dimensions in the Individuation -of Bodily Substances', Philosophical Studies (Ireland) 4 (1954), 60–79.

[29] 'Form and Matter', Philosophical Review 67 (1958), 499–521.

[30] W. Charlton, Aristotle's Physics, Oxford, 1970, p. 100. See also Moravcsik, op. cit., p. 7.

[31] Indeed, as we have just seen, the present material constituents of a living body are in general not identical with the materials from which the body came to be.

[32] See the Introduction, and the essays by Eslick and Bobik in The Concept of Matter, op. cit.

[33] Unless one wished to broaden the context of Aristotle's original discussion of the aitia of change, and propose genus as a way of "explaining" the constitution of a thing by noting that part of the nature of that thing it must have in common with a broader generic class.

[34] Op. cit., pp. 410–411.

[35] *Op. cit.*, p. 416.

[36] This is only approximately correct. See G. Stent, 'Explicit and Implicit Semantic Content of the Genetic Information', *Proceedings Fourth International Conference on the Unity of the Sciences*, New York, 1976, vol. 1, pp. 261–277.

[37] Again, this is, so far as we know, only very approximately true. See Stent, *op. cit.*

III

THEORY CHANGE

JOSEPH D. SNEED

DESCRIBING REVOLUTIONARY SCIENTIFIC CHANGE: A FORMAL APPROACH*

1. THE SCIENCE OF SCIENCE AND ITS PHILOSOPHICAL PROBLEMS

The term 'philosophy of science' covers a variety of activities – contributions to a variety of philosophical problems. Among these problems it has become common practice to distinguish between those related only to particular sciences and those related to scientific activity in general.

The first kind – hidden variable determinism in quantum mechanics, the role of 'mass' in classical mechanics, teleological concepts in evolutionary biology – are essentially problems of providing a clear, coherent conceptual framework for formulating the empirical claims of the particular sciences in question. The boundary between practicing this kind of philosophy of science and practicing the science itself is not always clear. In particular, solutions to these philosophical problems are bounded by empirical data in the same way as the theories themselves.

Philosophical problems relating to the nature of scientific activity in general – confirmation theory, characterizing scientific explanations and scientific theories in general – appear to be less clearly subject to constraint by any empirical facts. Literature about these problems is by no means unequivocal on this point. In much – especially the work in the logical empiricist tradition – there is a strong normative tenor.

I suggest that this 'duality' among problems in the philosophy of science is misleading. I maintain that philosophical problems about the nature of science-in-general are not, in any fundamental way, different from philosophical problems related only to particular sciences. In particular, there is no special sense in which the philosophy of science-in-general is a normative enterprise while the philosophy of particular sciences is not. Roughly, I maintain that there is an empirical, descriptive (but not *merely* descriptive) 'science of science'. Philosophy of

Butts and Hintikka (eds.), Historical and Philosophical Dimensions of Logic, Methodology and Philosophy of Science, 245–268.
Copyright © 1977 by D. Reidel Publishing Company, Dordrecht-Holland. All Rights Reserved.

science-in-general, in my view, deals with problems of providing a clear, coherent conceptual framework for formulating the empirical claims of specific theories of science.

The 'science of science' I have in mind is a social science. Its primary objects are, very roughly, groups of people – 'scientific communities' – engaged in a cooperative activity which produces, among other things, scientific theories. Scientific communities have properties – presumably related to the kind of products they produce – that differentiate them in interesting ways from other kinds of social groups. They interact in specific ways with the larger society. In the course of time, as a result of both internal and external factors, they come into being, fragment, coalesce and pass away. Likewise their products – 'scientific theories' – change and develop over time in ways intimately connected with the development of the communities that produce them. This is roughly the subject matter for a theory of science. To the extent that the products of scientific communities are of value to the larger society, a theory of science may have pragmatic implications. It may be of some interest to know whether the larger society can influence the 'output' of scientific communities and, if so, to what extent and by what means.

Traditionally the philosophy of science-in-general has not dealt with the full range of philosophical problems related to a theory of science. It has not dealt with questions like the identity conditions for 'scientific communities' nor with clarifying the concepts employed to describe community-related motivations of individuals in these groups. Rather, philosophy of science-in-general has focused on clarifying the concepts employed to describe some of the products of the scientific communities – roughly, scientific theories – the relations among them and their development over time. This narrow focus might be justified if, as a social science, the science of science has no philosophical problems not shared by a wide class of other social sciences, except those related to its products. I doubt that this is true.

I have spoken of the science of science and its conceptual foundations – the philosophy of science-in-general – as if they were completely void of normative implications. This is not so. But some distinctions are needed to reach a precise understanding of their normative implications.

First, an empirical theory of science, like any reliable empirical theory, has implications for action. It provides us with information we need to pursue some of our objectives. As a social science, the science of science may be particularly relevant to action since the 'social facts' it treats could possibly be modified by deliberate human action. In particular among the products of scientific communities are 'scientific facts' – systematic information about some range of phenomena. It is commonly believed that the most effective and efficient way to procure such information is to support (financially and otherwise) scientific communities. Indeed this is one way to formulate the claim that 'scientific activity is rational' – it is a good investment. Though commonly believed, this could nevertheless be false.

Whether or not the present system of scientific activity is nearly optimal (rational) in this economic sense and how it might be improved are among the things a successful theory of science might tell us. In this sense such a theory has clear normative implications. I believe Stegmüller may be getting at something like this when he notes:

... daß mit der Frage: "Was ist *rationales* wissenschaftliches Verhalten?" kein triviales, sodern ein schwieriges metatheoretisches Forschungsprogramm formuliert wird, in dessen Verlauf es sich als notwendig erweisen kann, gewisse sich zunächst anbietende Rationalitätskriterien später *als zu schablonenhaft* zu erkennen und sie zugunsten anderer Merkmale, die durch eine differenziertere Betrachtungsweise gewonnen worden sind, *zu verwerfen.* ([5] pp. 298–99.)

To this I have added only a politico-economic measure of 'rationality' and an explicit commitment to an empirical "metatheoretisches Forschungsprogramm".

But in the 'science of science' as in other empirical sciences, *mere* empiricism is not enough. Just as rational reconstruction of classical particle mechanics, classical equilibrium thermodynamics and elementary quantum mechanics provides valuable (so I judge) insight into the empirical content of these theories, so does rational reconstruction of specific theories in the science of science. I regard Kuhn's account ([2] and related articles) as one such specific theory. In Ch. VIII of [4] I have sketched a *partial* rational reconstruction of Kuhn's theory. Stegmüller (in [5] K.IX) has provided a much more detailed and insightful *partial* rational reconstruction of this same theory. Both these reconstructions are 'partial' in that they focus primarily on what Kuhn had said about the products of scientific communities – scientific theories and the empirical claims associated with them. Other aspects of his theory dealing with the scientific communities are however peripherally touched. In particular, both, Stegmüller and I, in somewhat different ways, try to explain what it is for a person "zu verfügen über" or 'to have' a theory.

2. THE PRODUCTS OF SCIENTIFIC COMMUNITIES

Perhaps the central feature of the partial reconstructions of Kuhn's theory of science that Stegmüller and I have offered is our concept of a scientific theory – a product of scientific community. Roughly speaking, in order to make sense out of what Kuhn was telling us about scientific activity we found it convenient to employ a concept of scientific theory somewhat different from that commonly used by philosophers of science-in-general.

Philosophers of science-in-general have commonly viewed scientific theories as sets of statements standing in certain logical relations with each other. For rather technical reasons something more than a set of statements is required to depict theories with sophisticated mathematical formalism and several interrelated applications. In addition, we found that the 'statement conception' was simply incompatible with many things Kuhn wanted to tell us about scientific theories. For example, one thing he wanted to tell us was that a necessary condition for membership in a specific scientific community was 'holding' a specific theory, but that members of the same scientific community both simultaneously and over time might differ significantly in their specific beliefs about the theory's subject matter. For this and many similar reasons (see [5] VII-7 and [4] VII) we adopted roughly the following concept of scientific theory. A scientific theory is a *conceptual structure* that can generate a variety of empirical claims about a loosely specified, but not completely unspecified, *range of applications.* Note that we do not get along without empirical claims. We just need another entity as well; roughly, a conceptual structure-cum-range of applications. That we choose to call this entity a 'scientific theory' is, of course, a matter of terminological convenience. That we found such an entity essential in reconstructing Kuhn's theory of science is not.

With the help of this concept of scientific theory we were able to reconstruct Kuhn's distinction between normal and revolutionary scientific change. Roughly, normal change is change in the body of empirical claims commonly held within the scientific community – holding the theory, in our sense, fixed. Revolutionary change is switching theories, or perhaps more accurately, the dying of a community holding one theory and the coming into being of another holding another 'competing' theory. Further, certain important relations among theories in our sense and their components may be defined – for example those of specialization, equivalence and reduction. This suggests, but surely does not prove, that interesting facts about how scientific theories develop over time be expressed as relations among scientific theories in our sense. In particular, Stegmüller has suggested that the reduction relation might be the appropriate one to capture the informal, intuitive notion of scientific progress.

I will consider Stegmüller's suggestion in more detail below. Here, I want to make more explicit the claim that the basic concepts of a theory of the *products* of science – a proper part of a full theory of science – are theories, in our sense, and *some* relations among them – not necessarily those just mentioned. Roughly, the claim is that everything interesting a scientist of science might want to say about the products of science can be said within this conceptual framework. This is obviously a programmatic claim – an injunction to try. Indeed, what I am sketching here is a 'core' for a theory of science.[1]

The program I have just sketched for a theory of the products of science is fraught with difficulties – some known to me and very likely many more as yet unknown to me. I want now to turn to some of the known difficulties – in particular the problem of formally distinguishing theory change that is intuitively progress from that which is not. In order to do this I must venture somewhat more deeply into the formal characterization of the concept of scientific theory.

3.   THEORIES AS SET-THEORETIC STRUCTURES

I will first describe a kind of set-theoretic structure – the 'core' of a scientific theory-element – and show how it may be used to make statements about another set-theoretic entity – the range of intended applications of the theory. Cores together with their applications are 'theory-elements' – the building blocks out of which rather complicated empirical claims may be constructed. 'Theories' in Kuhn's sense consist of a basic theory-element together with the means to construct a *variety* of 'nets' of theory-elements. Each net consists of a 'specialization' of the theory's basic element which corresponds to an empirical claim about the limited range of intended applications in the specialization. The entire theory-net corresponds to a rather complex empirical claim about the whole range of intended applications. Overlapping applications in various theory-elements connected by 'constraints' on theoretical concepts are the essence of this complexity. This complexity gives theoretical concepts a kind of 'concreteness' they would otherwise lack in providing 'determination methods' for them.

To proceed further we need some notation[2]

(D0) (A) $M \in \mathcal{M} =_{\text{def}} M$ is a non-empty set.[3]

  (B) For all $M, N, R, C, \bar{R} \in \mathcal{M}$;

  (1) $\text{Pot}(M) =_{\text{def}}$ the power-set of M;

  (2) $R : M \to N =_{\text{def}} R$ is a function with domain M and range N;

  (3) C is a *constraint* for M iff[4]

    (a) $\varnothing \notin C$

    (b) $C \subseteq \text{Pot}(M)$

    (c) for all $X, Y \neq \varnothing \in \text{Pot}(M)$, if $X \in C$ and $Y \subseteq X$, then $Y \in C$;

    (d) for all $x \in M$, $\{x\} \in C$;

  (4) If $R : M \to N$ then $\bar{R} : \text{Pot}(M) \to \text{Pot}(N)$ such that, for all $X \in \text{Pot}(M)$, $\bar{R}(X) =$ the image of X under R;

  (5) If $R \subseteq (M \times N)$ then $\check{R} =_{\text{def}}$ the converse of R;

  (6) R *reductively corresponds* M with N ($\text{rd}(R, M, N)$) iff

    (a) $R \subseteq (M \times N)$

    (b) $D_I(R) = M$

    (c) $\check{R} : D_{II}(R) \to M$

  (7) If $\text{rd}(R, M, N)$ then
  $\bar{R} =_{\text{def}} \{\langle X', X \rangle \in \text{Pot}(M) \times \text{Pot}(N) \mid \text{for} \quad \text{all} \quad x' \in X'$
  there is an $x \in X$ such that $\langle x', x \rangle \in R\}$

I next attempt to characterize as generally as I can the kinds of set-theoretic structures that appear in scientific theories.[5]

(D1) X is an $m+k$-*theory-element-matrix* iff
  (1) $X \in \mathcal{M}$;
  (2) m and k are integers: $0 < m$, $0 \le k$;
  (3)   For all $x \in X$, there exist $n_1, \ldots, n_m, t_1, \ldots, t_k \in \mathcal{M}$ such that $x = \langle n_1, \ldots, n_m, t_1, \ldots, t_k \rangle$.

The intuitive idea is simply that $x \in X$ is an $m + k$ tuple of sets, relations, functions, etc. The elements in x may 'overlap' in that some element may appear 'hidden' in another complex element as a domain or range. For example, in classical particle mechanics members of X will be of the form $x = \langle P, T, s, m, f \rangle$ where P is a set of particles, T a real interval, $s: P \times T \to R^3$ a C^2-function, $m: P \to R^+$ and $f: N \times P \times T \to R^3$ a C^2-function. It is not necessary to specify which elements of $x \in X$ are 'real' sets and which are relations so we may call them generally '*components*' of x. That such generality is needed is suggested by Moulines' treatment ([3]) of the concept of equilibrium states in classical equilibrium thermodynamics as a 'theoretical' component in the sense of the next definition. We will however frequently be speaking of examples of components which are numerical functions.

Next I define the formal core of a theory-element. This is the apparatus that is used to make empirical claims about the elements range of intended applications.

(D2) X is a *theory-element-core* iff there exist M_p, M_{pp}, M, C, m and k such that:
  (1) $X = \langle M_p, M_{pp}, M, C \rangle$;[6]
  (2) M_p is an $m + k$ theory-element matrix;
  (3) $M_{pp} = \{ \langle n_1, \ldots, n_m \rangle \mid \langle n_1, \ldots, n_m, t_1, \ldots, t_k \rangle \in M_p \}$
  (4) $M \subseteq M_p$;
  (5) C is a constraint for M_p.

There are three intuitive ideas compressed into this definition. First, there is a distinction between theoretical and non-theoretical components. Roughly, M_p is the set of all possible models of the *full* conceptual apparatus of the theory including theoretical components, while M_{pp} is the set of all models obtained by simply 'lopping-off' the theoretical components leaving only the non-theoretical part of the conceptual apparatus. I have called these respectively 'partial models' and 'potential partial models'. Second, there is the idea of laws formulated with theoretical components. This is captured by M which picks out the set of all possible models of the full conceptual apparatus just those which satisfy certain laws. Third, there is the idea that different applications of the theory-element are interdependent in the sense one may not – at least for theoretical functions – employ values of function in one application of the theory without regard for some of the values of the same function in other applications. This is captured by the constraints C on M_p. These have the effect of ruling out certain combinations of theoretical functionvalues in different applications. The intuitive idea is that a distinction may be drawn between what is ruled out by the structure of the theory's models M and what is ruled out by restriction on the way that structure is applied 'across' a number of different applications $C.^{7}$

That a theory-element core has these intuitive properties can only be seen by looking at how it is used to make empirical claims. The intuitive idea is simple. Given a theory-element core $K = \langle M_p, M_{pp}, M, C \rangle$ a class of sub-sets of M_{pp} (that is, a sub-set of Pot (M_{pp})) will be selected. Call this class $A(K)$. The selection rule is this. A subset of M_{pp} is in $A(K)$ if and only if theoretical components can be added to each member of it in a way that yields a subset of M (satisfies the theoretical laws) and such that the whole array of theoretical components satisfies the constraints C.

The definition of $A(K)$ requires some additional notation. The r-function of (D3-A) simply 'lops-off' the theoretical components from members of M_p. $A(K)$ is then the image of Pot $(M) \cap C$ under $\bar{r} - r$ extended twice in the manner of (D0-B-4) to yield a function $\bar{r}$: Pot (Pot (M_p)) → Pot (Pot (M_{pp})). It is convenient here also to introduce some notation needed for defining reduction relations. If R is a relation between the full theoretical structures of two theory-element cores, then $\hat{R}$ is the non-theoretical image of R (D3-C)).

(D3) If $K = \langle M_p, M_{pp}, M, C \rangle$ and $K' = \langle M_p, M_{pp}, M, C \rangle$ are theory-element-cores then:

(A) $r: M_p \rightarrow M_{pp}$ is such that $r(n_1, ..., n_m, t_1, ..., t_k) =_{\text{def}} \langle n_1, ..., n_m \rangle$

(B) $A(K) =_{\text{def}} \bar{r}(\text{Pot } (M) \cap C)$

(C)   if $R \subseteq M'_p \times M_p$ then

$$\hat{R} = \{\langle x', x \rangle \in M'_{pp} \times M_{pp} | \text{there is a } \langle y', y \rangle \in R \text{ and}$$
$$x' = r'(y') \text{ and } x = r(y)\}[8]$$

We can now characterize the concept of a theory-element and begin to see how a theory is related to empirical claims.

(D4) X is a *theory-element* only if there exist[9] K and I such that:
   (1) $X = \langle K, I \rangle$;
   (2) $K = \langle M_p, M_{pp}, M, C \rangle$ is a theory-element core;
   (3) $I \subseteq M_{pp}$.

First, note that this is only a necessary condition for a theory-element. The set I is to be interpreted as the range of intended applications of the element – what the theory-element is about. The only requirement (D4) puts on I is that its members have the structure characteristic of the non-theoretical part of K – that they be members of M_{pp}.[10]

I now define the claim of the element $\langle K, I \rangle$. It is just the claim that the range of intended applications I is among those sub-sets of M_{pp} that are singled out by K – that is that $I \in A(K)$.

(D5)   The *claim* of the theory-element $\langle K, I \rangle$ is that $I \in A(K)$.

The idea here is again simple. M_{pp} is all possible non-theoretical descriptions of some body of phenomena; I consists of descriptions of phenomena that actually occur. The theory-element core K narrows Pot (M_{pp}) down to $A(K)$ – it restricts the range of possibilities. It claims to do this in a way that narrows down onto I – what is actually observed to occur.

The claim of a theory-element $\langle K, I \rangle$ may well be a rather crude narrowing down of the range of non-theoretically described possibilities. Indeed, it may be no narrowing down at all. That is, it may be that $A(K) = M_{pp}$ and the claim of $\langle K, I \rangle$ is vacuous.[11] But generally the claims of single theory-elements – even if non-vacuous – are not the only claims associated with theories. Typically, it is claimed that various special theoretical (and perhaps non-theoretical) laws hold in various sub-sets of some basic range of applications of I. Perhaps these special laws also have special constraints associated with them. It is roughly the claim of some basic theory-element plus *some* array of special claims like this that constitute the empirical content of scientific theories.

I attempted in [4] to deal with special laws and constraints using what I called an 'expanded core' for a theory. I now believe there to be a somewhat more perspicuous, but equivalent way of describing these features of theories.[12] First we define the notion of a 'specialization' of a theory-core. Intuitively, a specialization of $K = \langle M_p, M_{pp}, M, C \rangle$ assigns to some sub-set of M_{pp} certain special laws represented as further restrictions on the set M together perhaps with some constraints associated with these laws – represented as further restrictions on C.[13]

(D6) If T' and T are theory-elements then T' is a *specialization* of T iff

    (1) $M'_{pp} \subseteq M_{pp}, \quad M'_{pp} \in \mathcal{M}$

    (2) $\mathrm{Pot}\,(M'_{pp}) \cap A(K) \neq \varnothing$

    (3) $M'_p = \{x \mid x \in M_p \text{ and } r(x) \in M'_{pp}\}$

    (4) $M' \subseteq M$

    (5) $C' \subseteq C$

    (6) $I' = I \cap M'_{pp}$[14]

The reason for requiring $\mathrm{Pot}\,(M'_{pp}) \cap A(K) \neq \varnothing$ (D6-2) is simply that there would be no point in postulating special laws to hold in situations that have already been ruled out by the initial K. (D6-3) just lets the choice of M'_{pp} determine M'_p in a way that assures that T' is a theory element. Requiring $I' = I \cap M'_{pp}$ means that in choosing M'_{pp} we intend that *everything* in I having this structure is an intended application of the specialization. This can be weakened, but doing so prevents us from regarding 'core-nets' ((D11) below) as determining sub-sets of some initial M_{pp} in the same way that single cores do ((D12) below). For details see [1].

Next we define a 'theory-net' as a set of theory-elements together with the specialization relation.[15] We also require that cores appearing in the net be uniquely associated with their intended applications (D7-4).

(D7) N is a *theory-net* iff there exist $|N|$ and $\leqslant$ such that:

    (1) $N = \langle |N|, \leqslant \rangle$;

    (2) $|N| \in \mathcal{M}$ is a finite set of theory-elements;

(3) $\leq$ is the specialization relation on $|N|$;
(4)   for all $\langle K, I \rangle$, $\langle K', I' \rangle \in |N|$, $K = K'$ iff $I = I'$.

Theory-nets are partially ordered sets (T1-A) and the sub-sets of M_{pp} picked out by specialization-related theory-elements have the properties that 'specialization' suggests (T1-B). That is $A(K)$'s determined by specializations of K' are generally smaller than $A(K')$. They narrow down the range of possibilities permitting us to make stronger claims.

(T1)   If N is a theory-net then:
   (A) N is a partially ordered set;
   (B)   if $T, T' \in |N|$ are such that $T \leq T'$ then:
      (1) $A(K) \subseteq A(K')$,
      (2) $I \subseteq I'$,
      (3)   if $I \in A(K)$ then $I \in A(K')$.

Theory-nets are pretty general. Examples of practical interest appear to be quite special cases. We shall be mainly concerned with theory-nets which have a single 'basic theory-element' at the top of the net.

To specify the properties of theory-nets that are of interest to us we need some notation. We define the 'basic set' of a net as the set of its upper bounds ((D8-A)), and the set of all intended applications appearing anywhere in the net N, $Q(N)$ ((D8-B) and (-C)).

(D8)   If $N = \langle |N|, \leq \rangle$ is a theory-net then:
   (A) $\mathcal{B}(N) =_{\text{def}} \{ x \in |N| \,|\, \text{for all } z \in |N|, \text{ if } x \leq z \text{ then } x = z \}$;
   (B) $\bar{Q}(N) =_{\text{def}} \{ I \,|\, \text{there is a } \langle K', I' \rangle \in |N| - \mathcal{B}(N)$
                                 such that $I = I' \}$.
   (C) $Q(N) =_{\text{def}} \bigcup_{I' \in Q(N)} I$

Not only do we need to talk about theory-nets, we must also talk about 'parts of theory-nets'. To this end we introduce the notations of 'expansion' and 'initial part' ((D9)). N is an expansion of N' iff N' is contained in N. N' is an initial part of N iff it comprises the 'top' of N.

(D9)   If N and N' are theory-nets then:
   (A) N is an *expansion* of $N'(N' \subseteq N)$ iff:
      (1) $|N'| \subseteq |N|$;
      (2) $\leq' = (\leq \cap |N'| \times |N'|)$.

(B) N is an *initial part* of N' iff:
(1) $N' \subseteq N$;
(2)   for all $x \in (|N| - |N'|)$, $y \in |N'|$ not $y \leq x$.

Theory-nets have 'core-nets' associated with them in a natural way (D10) and these core-nets are isomorphic to the theory-nets ((T2)).[16]

(D10)   If $N = \langle |N|, \leq \rangle$ is a theory-net then $N^* = \langle |N^*|, \leq^* \rangle$ is the *core-net induced by* N iff
(1) $|N^*| = \{K \mid \langle K, I \rangle \in |N|\}$,
(2) $\leq^* = \{\langle K, K' \rangle \in (|N^*| \times |N^*|) \mid \langle K, I \rangle \leq \langle K', I' \rangle\}$

(T2)   If N is a theory-net then N is a partially ordered set and N^* is isomorphic to N under $\theta: |N| \leftrightarrow |N^*|$ such that $\theta(K, I) = K$.

Just as the theory-core K picks out a sub-set of Pot (M_{pp}), so does a theory-core-net under the core K. Intuitively, N^* picks out just those members of Pot (M_{pp}) whose elements may be supplied with theoretical functions in ways which satisfy special laws and constraints in applications where they are postulated to apply, as well as those laws and constraints which are postulated to apply, in applications standing above them in the net – including, of course, those of K. The formal definition follows.

(D11)   If N is a theory-net such that $\mathcal{B}(N) = \{\langle K, I \rangle\}$ then: $A(N^*) = \{X \leq M_{pp} \mid$ for all $K' \in |N^*|$, $X \cap M'_{pp} \in A(K')\}$.[17]

We can now give a more complete account of the relation between a basic theory-core K and empirical statements about a range of intended applications I. For some theory-net N^* based on $\langle K, I \rangle$ it is claimed that $I \in A(N^*)$. This claim entails that some sub-sets of I are in $A(K')$'s determined by K's in the net. This suggest that one may look at the claim $I \in A(N^*)$ as an array of claims of the form $I' \in A(K')$ associated with sub-theories $\langle K', I' \rangle$ under $\langle K, I \rangle$. This array of sub-theories is the 'theory-net based on $\langle K, I \rangle$'. This is summarized formally in the following theorem.

(T3)   If N is a theory-net such that $\mathcal{B}(N) = \{\langle K, I \rangle\}$ then $I \in A(N^*)$ for all $\langle K', I' \rangle \in |N|$, $I' \in A(K')$.

This account of a theory and related empirical statements permits us to deal with situations in which special laws and constraints are claimed to hold in certain special applications, but not in all. It shows how a rather weak, perhaps even vacuous, theory-core may be supplemented to strongly restrict the range of possibilities. Moreover, it suggests a way of depicting the temporal development of a theory.

First, consider a theory-element $\langle K, I \rangle$ in which the range of applications I is regarded as extensionally described. The historical development of a theory might be represented as successive attempts to construct theory-core-nets N_i^* based on K so that $I \in A(N_{i+1}^*) \subset A(N_i^*)$. That is, the top of the theory-net remains fixed while the rest changes through time in an effort to get an ever more precise 'fix' on I. This 'zeroing-in' on I need not proceed uniformly. One could start to extend the net in one direction; see a better way; back-up and start in a different direction.

This models a situation in which *only* the claim of the basic theory-element $\langle K, I \rangle$ and roughly the 'commitment' to use nets under K in trying to say more about I remain stable. But sub-nets could as well be held 'fixed' in a similar way. That is there could be a $K' \leqslant^* K$ such that $\langle K', M'_{pp} \cap I \rangle$ appeared in every theory-net based on $\langle K, I \rangle$ throughout the theory's development. This would indicate roughly that practitioners of the theory were not simply committed to using nets under K in dealing with I, but also in handling some sub-sets of I in a particular way.

Interesting examples of theories with extensionally described intended applications do not abound. More typically, the range of intended applications I is described intensionally with some $I_t \subseteq I$ – the members of I discovered up-to-t – described extensionally. For fixed I_t the theory based on $\langle K, I \rangle$ may develop as just described. But this theory may also develop by enlarging I_t to $I_{t'}$ and finding some theory-core-net N under K so that $I_{t'} \in A(N)$. That is, more of the intended applications may be brought within the scope of what the theory says about extensionally described applications. Roughly, the theory based on $\langle K, I \rangle$ may develop both intensively (more narrowly specifying I_t) and extensively (enlarging I_t) by extending and shifting the theory-net under $\langle K, I \rangle$.

Among theories with intensionally described intended applications, those whose intended applications described by 'paradigm examples' have received particular attention. Kuhn has suggested that description of intended applications by paradigm examples may be the rule for 'interesting' scientific theories. This possibility may be represented in the theory-net formalism in the following way. A paradigm set of extensionally described applications I_o has associated with it a 'paradigm-theory-net' N_p. This paradigm theory-net determines the development of the theory in two ways. First, I is described roughly as all those things sufficiently like I_o. (This can be made somewhat more precise. For example see [4] p. 286ff. and [5] p. 148ff., 218ff., 225ff.). Second, the paradigm theory-net must be a sub-net of any larger net appearing during the theory's development. That is, certain subtheories and relations among them 'anchor' the development of the theory.

These intuitive ideas are captured in (D13). T_b is the basic theory-element consisting of K_b – the basic conceptual structure of the theory – and I – the full range of intended applications. I_o is the sub-set of paradigm applications and N_p is the paradigm theory-net which applies K_b to I_o – perhaps with the help of specializations of K_b. $\mathcal{N}$ is the set of all nets which are expansions of N_p. One might want to require that N_p be an initial part of all members of $\mathcal{N}$. This would rule out tampering with the paradigm net by squeezing new applications in between the old ones. Whether or not this ever happens is not clear to me.

(D12) $\mathcal{T}_k$ is a *Kuhn-theory* iff there exist T_b, I_o, N_p, $\mathcal{N}$ such that:

    (1) $I_k = \langle T_b, I_o, N_p, \mathcal{N} \rangle$;

    (2) $T_b = \langle K_b, I \rangle$ is a theory-element such that $I_o \in I$;

    (3) N_p is a theory-net such that:

        (a) $\mathcal{B}(N_p) = \{T_b\}$;

        (b) $Q(N_p) = I_o$

    (4) $\mathcal{N} = \{N \,|\, \text{(a)} \; N \text{ is a theory-net}$

        (b) $\mathcal{B}(N) = \{T_b\}$

        (c) $N_p \subseteq N\}$

Kuhn-theories remain stable while the empirical claims associated with them are relative to the theory-net one chooses from $\mathcal{N}$. This is made precise by (D13).

(D13) If $\mathcal{T}_k = \langle T_p, I_o, N_p, \mathcal{N} \rangle$ is a Kuhn-theory $T_p = \langle K, I \rangle$ and $N \in \mathcal{N}$ then the *claim of J_k with respect to* N is that $I \in A(N^*)$.

Two possibilities for 'normal' scientific progress within the framework of a Kuhn-theory are depicted by (T4). If N_1 and N_2 are theory-nets associated with a Kuhn-theory and N_2 is an expansion of N_1 then N_2 may explicitly treat more members of I than N_1 or N_2^* may say something sharper about I than N_1^*. A third possibility for normal scientific progress is that $A(N_1^*) = A(N_2^*)$ but N_2^* is 'neater' or 'more elegant' than N_1^*. Important though this form of progress may be (to the careers of scientists) it appears to escape formal representation.

(T4)   If $\mathcal{T}_k$ is a Kuhn-theory and $N_1, N_2' \in \mathcal{N}$ such that $N_1 \subset N_2$ then
    (A) $I_o \subseteq Q(N_1) \subseteq Q(N_2) \subseteq I$
    (B) $A(N_2^*) \subseteq A(N_1^*)$.

Though I have used the theory-net formalism in an attempt to partially reconstruct Kuhn's theory of science, the formalism is independent of his theory and flexible enough to accommodate many other views. For example, one could simply drop the element I_o and N_p from the theory-tuple to obtain a conception of scientific theory that had nothing to do with Kuhn's concept of paradigm. On the other hand, the formalism may be extended in the other direction as well.

Paradigm theory-nets may come into being with the theory's basic formalism – the basic theory-core K_b – and remain unchanged throughout the theory's development. This, I believe, was Kuhn's initial conception. But, the theory-net formalism does not wed us to this conception. Theory-core-nets and theory-nets are homogeneous.[18] Anything we say about the whole net can, in principal, be said about any sub-net. There can be 'little paradigms' for sub-nets which come into being and pass away during the development of the 'overarching' theory based on $\langle K, I \rangle$.[19] Moreover, there could be a shift over time in the 'anchor structure' N_p of the theory-nets without necessarily changing I.

I do not believe these exhaust the interesting descriptive possibilities for the theory-net formalism in 'normal science'. But let us move directly to its possible application to 'revolutionary' scientific change.

4. REDUCTION RELATIONS

I want now to suggest how the formalism developed above may be used to describe some interesting relations among scientific theories. I w.ll be trying to get at relations that capture our intuitive idea that a theory $\mathcal{T}$ is more powerful than $\mathcal{T}'$ in the sense that $\mathcal{T}$ 'could explain' everything that $\mathcal{T}'$ 'explains' and perhaps more. These are examples of the kinds of relations that one might look for (but perhaps not find) between theories historically separated by a 'scientific revolution'.

What I think emerges from this investigation is the realization that our gross intuitive conception of what counts as 'scientific progress' may be sharpened in a number of quite different ways. Taking account of the full complexity of the kinds of things that scientific theories are, one can distinguish a number of relations among theories that might plausibly indicate 'scientific progress'. This suggests, at least, that a discussion of 'whether science progresses' in anything like the intuitive sense of 'progress' should be accompanied by a patient examination of the relations that actually appear to hold between putative examples of theories historically separated by 'scientific revolutions'.

The most general relation here is that of one theory-element's reducing to another. It may (but need not) hold between theories elements completely unrelated conceptual apparatus (different theory-element cores). It may be used to define other interesting relations among theory-elements, theory-nets and theories. The intuitive idea of theory-element reduction is this. Every application of the reduced theory-element corresponds to at least one application of the reducing theory-element and what the reduced theory-element says about a given application is entailed by what the reducing theory-element says about any corresponding application.

Notation needed to make this precise is provided in (D0-6-7). A reductive correspondence R between N' and N is simply a one-many relation from N' to N. The associated R simply extends R to the power sets of N' and N. We need the latter because, in our view, theories generally make claims about whole sets of applications rather than single application. We now define a weak reduction relation between theory-elements. We say R weakly reduces T' to T just when R is a reductive correspondence between the non-theoretical structures in K' and K and $X \in A(K)$ assures that a R-corresponding X' is in $A(K')$. These are pure formal requirements. In contrast, (D14-3) requires that the ranges of application be such that every application of T' R-correspond to some application of T.

(D14)   If T' and T are theory-elements then R *weakly reduces* T' to
$T(\overline{RD}(R, T', T))$ iff
(1)   rd (R, M'_{pp}, M_{pp})
(2)   for all $\langle X', X \rangle \in \mathrm{Pot}\,(M'_{pp}) \times \mathrm{Pot}\,(M_{pp})$, if $X \in A(K)$ and
$\langle X', X \rangle \in \tilde{R}$ then $X' \in A(K')$.
(3) $\langle I', I \rangle \in \tilde{R}$.

The weak reduction relation R just defined may be regarded as
capturing the bare essentials of a reduction relation between two
theories in which T' and T are the basic theory-elements standing at
the top of all theory-nets of the respective theories. Roughly, that such
a weak reduction relation between the basic theory-elements exists is a
necessary condition for there to be a reduction relation between the
theories.

Weak reduction is a correspondence between the non-theoretical
parts of theory-element matrices taken as unanalyzed wholes. It may
adequately represent at least some cases of reduction where the
theoretical concepts of two theories are so different as to be 'incompar-
able'. This might be this case, for example in theories separated by a
'scientific revolution'. Weak reduction requires only that non-
theoretical structures of the two theories be 'comparable' and this
'comparability' need not be between individual components of the
matrices. It is hard to imagine a weaker kind of 'comparability'.

But there are two intuitive grounds for regarding weak reduction as
unsatisfactory in many cases. First, some 'interesting' reduction relations appear to have to do with the theoretical components of matrices. Second, they appear to involve a 'piecewise' correspondence
between the components of the matrices. For example, in reducing
rigid body mechanics to particle mechanics 'moment of inertia' should
correspond to 'mass'. The first intuitive requirement we may meet by
requiring the reduction relation to be a reductive correspondence
between M'_p and M_p. The second we may partially meet by requiring
that this correspondence be 'separable' into a non-theoretical and
theoretical part. In this way one may define a strong reduction relation
between theory-elements (see [4] p. 216ff.). But for present purposes
we may avoid the formal definition and speak simply of 'reduction' –
without specifying 'strong' or 'weak'.[20]

Let us now consider the 'preservation properties' of reduction relations. One might suspect that, once a reduction relation had been discovered between theory-elements, it could be preserved in the following sense. If R reduces to T' to T and T'_1 is some specialization of T' then there will always be some specialization T_1 of T such that R reduces T'_1 to T_1. Roughly, it should always be possible to replicate the effect of any special laws in the reduced theory by corresponding special laws in the reducing theory.

For both weak reduction and strong reduction this is so, provided one is not at all fastidious about what counts as a special law of the reducing theory. For weak reduction, the R-image of $A(K'_1)$ can always be used to construct the requisite K_1.

These facts are summarized in the following theorem.

(T6) If T' and T are theory-elements and $RD(R, T', T)$ then for all T'_1 such that T'_1 is a specialization of T' there is some T_1 such that T_1 is a specialization of T and $RD(R_1, T'_1, T_1)$ where $R_1 = R \,|\, (M'_p)_1 \times (M'_p)_1$.

These facts about preservation of reduction relations through specialization suggest that for theories based on $\langle K', I' \rangle$ and $\langle K, I \rangle$ reduction is completely captured by a reduction relation between the basic theory-elements. But this ignores the intuitive requirement that we can not be completely undiscriminating in choosing refinements of $\langle K, I \rangle$ which 'reduce' given refinements of $\langle K', I' \rangle$, roughly, we can not use arbitrarily contrived special laws of $\langle K, I \rangle$ to reproduce the effect of special laws in $\langle K', I' \rangle$; we must use special laws of $\langle K, I \rangle$ that already have some 'standing' within the latter theory.

One way to handle this intuitive requirement is simply to regard reduction as a relation between specific theory-nets associated with the theories. Roughly, the theory-net reduction relation must reduce the basic theory element T' to the basic theory-element T and every theory-element of N' to some theory-element of N.[21] The following definition makes this precise.

(D15) If N' and N are theory-nets T' and T theory-elements such that $\mathscr{B}(N') = \{T'\}$ and $\mathscr{B}(N) = \{T\}$ then R reduces N' to N $(^0R^0D(R, N', N))$ iff:
(1) $RD(R, T', T)$

(2) for all $T_1' \in N'$ there exist some $T_1 \in N$ such that $\overline{RD}(R_1, T_1', T_1)$ where $R_1 = R \mid (M_p')_1 \times (M_p)_1$

The concept of reduction between theory-nets allows us to distinguish four senses of 'reduction' between Kuhn-theories (D16). Each of these may be duplicated for 'weak reduction'.

(D16) If $\mathcal{T}_k'$ and $\mathcal{T}_k$ are Kuhn-theories and $\langle N', N \rangle \in \mathcal{N}' \times \mathcal{N}$ then:

    (A) R *basic-reduces* $\mathcal{T}_k'$ *to* $\mathcal{T}_k$ iff $\overline{RD}(R, T_b', T_b)$;

    (B) R *reduces* $\mathcal{T}_k'$ *to* $\mathcal{T}_k$ *relative to* $\langle N', N \rangle$ iff ${}^0\overline{R}{}^0D(R, N', N)$;

    (C) R *prospectively reduces* $\mathcal{T}_k'$ *to* $\mathcal{T}_k$ *relative to* $\langle N', N \rangle$ iff

        (1) R reduces $\mathcal{T}_k'$ to $\mathcal{T}_k$ relative to $\langle N', N \rangle$

        (2) for all N_1', $M' \subseteq N_1'$ there exists an N_1, $N \subseteq N_1$ such that ${}^0\overline{R}{}^0D(R, N_1', N_1)$;

    (D) R *completely reduces* $\mathcal{T}_k'$ *to* $\mathcal{T}_k$ iff for all $\langle N', N \rangle \in \mathcal{N}' \times \mathcal{N}$ ${}^0\overline{R}{}^0D(R, N', N)$.

'Basic reduction' ((D16-A)) is a simple reduction between the basic theory elements. Basic reduction alone appears to be too weak to capture our intuitive idea of reduction. This is evident when one notes that claims associated with basic theory-elements may be trivial in that $A(K_b) = M_{pp}$ for non-trivial theories. Were this case $\mathcal{T}_k'$ then any reductive correspondence between M_{pp}' and M_{pp} would reduce $\mathcal{T}_k'$ to $\mathcal{T}_k$.

A more satisfactory reduction concept for Kuhn-theories is provided by 'reduction relative to specific theory-nets' ((D16-B)).

This relation reflects the intuitive ideas that let us define reduction between theory-nets. But there is some intuitive reason to think that reduction of theories relative to specific theory-nets is not the relation we are interested in. Many theory-nets (in the course of time) may be associated with the same theory. Intuitively, we think that reduction is a time-independent relation holding between the theories themselves. A possible move to obtain a time independent reduction relation is to require 'complete reduction' ((D16-C)) – *every* pair of theory-nets reduced by R. But complete reduction is clearly too strong. Big J_k'-nets will never reduce to little J_k-nets.

There is the illusory appearance of a happy compromise between net-relative reduction and complete reduction – 'prospective reduction relative to specific nets' ((D16-C)). The idea is that R reduces N' to N

and for every expansion of N' there is some expansion of N such that R reduces the expansions. One may think of the initial $\langle N', N \rangle$ either as the paradigm-net pair $\langle N'_p, N_p \rangle$ or as a pair $\langle N'_t, N_t \rangle$ depicting the state of both theories as some moment in their historical development. For paradigm-nets as the initial pair prospective reduction means that R reduces the paradigm-nets and, for every N' and $\mathcal{N}'$ there is some N in $\mathcal{N}$ such that R reduces N' to N. But, as a consequence of (T6)/(T7-B) prospective reduction relative to $\langle N', N \rangle$ adds nothing to reduction relative to $\langle N', N \rangle$. Intuitively, so long as we may arbitrarily expand N we can always find an expansion to which R will reduce a given expansion of N'. (T7-A) makes explicit the fact that basic reduction is entailed by all stronger reduction relations.

(T7)   If $\mathcal{T}'_k$ and $\mathcal{T}_k$ are Kuhn-theories and $\langle N', N \rangle \in \mathcal{N}' \times \mathcal{N}$ then:
   (A)   if R reduces $\mathcal{T}'_k$ to $\mathcal{T}_k$ relative to $\langle N', N \rangle$ then R basic-reduces $\mathcal{T}'_k$ to $\mathcal{T}_k$;
   (B) R reduces $\mathcal{T}'_k$ to $\mathcal{T}_k$ relative to $\langle N', N \rangle$ iff R prospectively reduces $\mathcal{T}k$ to $\mathcal{T}'$ relative to $\langle N', N \rangle$.

These facts suggest that we must either accept some net-relative reduction as the appropriate reduction between Kuhn-theories or strengthen the concept of specialization in some way that blocks the result of (T6). The latter possibility essentially amounts to specifying a class of 'admissible' or 'law-like' sub-sets of M_b – the laws in the basic theory element. Specializations would then be restricted to those containing admissible special laws. The analogue of (T6) would then be generally false, though it might hold for some specific sets admissible in special laws.

On the other hand net-relative reduction may not be so objectionable as it first appears. To see this we must make a 'pragmatic' distinction between two kinds of reduction. First, there is a situation in which the reducing theory is developed (temporally) before the reduced theory and more or less explicitly intended as an 'off-shoot' of the reducing theory.

Classical particle mechanics has roughly this relation to classical rigid body mechanics and hydrodynamics. Initially one expects the 'off-shoot' theories to reduce to the 'parent' theory in the sense that the paradigm net of the off-shoot theory reduces to the net of the parent theory that depicts this theory's development at the 'birth date' of the off-shoot. But, once conceived, the offshoot theory has a life of its own. Applications for it may be found that go beyond those reducible to the initial net of the parent theory as well as expansions of this net obtained through 'natural' development of the parent theory. But the relation between these theories may be so close that any net of the parent theory produced as an 'image' of a net of the offshoot theory would be acceptable as a 'natural' development of the parent theory. Roughly, the temporal development of offshoot and parent run parallel. Here net-reduction is acceptable because, even if we could be precise about them, stronger concepts of specialization would be irrelevant. Roughly, the stronger specialization concepts for off-shoot and parent would be isomorphic.

In contrast, consider the case in which the (putative) reducing theory develops later than the reduced theory – the case of scientific revolutions. At the birth date of the 'reducing' theory its proponents have only its paradigm net and the hope that expansions of this net can be found. The proponents of the 'reduced' theory have some extensive expansion of their paradigm net. In general, not even an 'overlapping' of these two nets need occur initially. Even when there is some initial overlapping, reduction is not to be expected. What is perhaps to be expected is that the development of the new theory will ultimately overtake that of the old. That is, at some time the net N'_t associated with the old theory will reduce to N_t associated with the new theory, and not conversely. (Note that N'_p need not necessarily reduce to N_p.) Such a development would clearly count as 'progress', but not 'ulti-mate defeat' for the old theory. There is nothing in the 'facts' of the situation to prevent the old theory from developing further in direc-tions that can not be reduced to the net N_t – though they could be reduced to *ad hoc* expansions of N_t.

Here 'having a reduction relation' between two theories is rather like 'having a theory' (see [4] pp. 265–8 and [5] pp. 189–95). It is an essentially pragmatic concept. One has discovered a relation R that reduces the net N'_t to the net N_t. One *believes*, perhaps with good reason, that R will suffice to reduce all nets arising out of the 'natural' development of T' to nets arising out of the temporally parallel 'natural' development of T. But this belief may not be grounded on conclusive proof. These considerations suggest that perhaps reduction of theories relative to specific nets is all we need to deal with the question of 'scientific progress'. They suggest that we do not need a concept of theory in which the possibilities for its future development are so strongly specified (through a strong specialization concept) that it can be established 'once-and-for-all' that one theory can not keep up with another. Though we do not *need* such a concept to rationally allocate our research resources, it would still be nice to have. It would reduce the uncertainty in our allocation problem – something that is always nice and sometimes purchasable at acceptable cost. Moreover, it could be the case that some theories have such limitations 'built into' them while other don't. This suggests that it makes sense to look for strong restrictions on a theory's development – like perhaps invariance principles in mechanics – but that we can get on without them.

University of Munich

NOTES

* A version of this paper was published in advance in *Erkenntnis* **10** (1976), 115–146.

[1] I will not pursue this formal line here. I only remark that, had I realized at some time that my discussion of Kuhn's theory was a logical reconstruction of an empirical theory of science, I would have been tempted to try casting it in the same framework I suggested for physical theories. Perhaps fortunately for my readers, my insight was not so sharp.

[2] Here and in what follows I rely heavily on the notation of Stegmüller [5].

[3] I use 'set' to denote both naive sets as well as sets of ordered tuples. So in some cases '$M \in \mathcal{M}$' may mean M is a relation or function.

[4] The definition of 'constraint' here differs from that in [3] p. 170 and [4] p. 128 in respects. First the null-set is excluded from (C) (D0-B-3-a) and finally a 'transitivity' requirement (D0-B-3-c) is imposed. The former is a technical nicety required to make (D6-2) intuitively meaningful. The latter is required in the proof in [4] p. 282. Because of (D6-6) this condition is not required for any theorems mentioned here. But weakening (D6-6) to $I' \subseteq M'_{pp} \cap I$ would require us to use this condition in establishing a partial analogue of (T3).

[5] The following definition is a generalization of (D24) ([4], p. 161) and the corresponding (D1) ([5], p. 123) which eliminates one of the differences mentioned in the discussions accompanying the earlier definitions – namely, the restriction of the nature of the relations in the matrix.

[6] The r-function appearing in the definition of a 'core for a theory of mathematical physics' in [4], p. 171 and [5], p. 128 can now be defined in terms of M_p and M_{pp} (see (D3) below) and hence need not appear explicitly. The earlier treatment did not permit this because it avoided hanging anything on the less satisfactory definition of 'matrix'.

[7] Actually (D2) is so generally formulated that constraints may apply to non-theoretical as well as theoretical components. Whether such generality is actually needed is an interesting question. (D2) may easily be modified to restrict the constraints to theoretical components only, but this complication is not essential here.

[8] Here 'r' denotes the function $M'_p \rightarrow M'_{pp}$ defined by (D3-1). Generally, we adopt the convention that components and other sets associated with indexed K's will be decorated with the same index. For example, 'C_i' denotes the constraints in K_i.

[9] Theory-elements correspond to theories of mathematical physics in [4] and [5]. The present terminology emphasizes the fact that *theories* are generally more complex entities than single theory-elements.

[10] It is possible to say more in the direction of sufficient conditions. Some tentative steps in this direction were made in [4], p. 260 and [5], p. 189, but I shall not venture into these matters here.

[11] This turns out to be the case in one plausible formulation of classical particle mechanics. See [4], p. 118 and [5], p. 110.

[12] The basic idea of this alternative treatment – the specialization relation – was suggested to me by W. Balser. For a more detailed and extensive treatment of inter-theoretical relations in this manner see [1].

[13] A specialization of T is itself a theory-element but it plays the role of the 'application function' α in earlier treatments [4], p. 179 and [5], p. 131.

[14] Note that (D6-4) and (-5) together with the requirement that T' be a theory element insure that the special laws M' and special constraints C' actually hold for some members of the M'_p determined by M'_{pp} – that is that $A(K')$ is not void. This corresponds to (D9-5-c) in [5], p. 131 – a condition that was lacking in defining the application relation in [4], p. 181. The counterpart to (D9-5-b) in [5] is (D6-6) which effectively requires that nested sequences of application sets be associated with nested sequences of laws.

[15] Theory-nets are analogous to applied cores for theories of mathematical physics in [4], p. 181.

[16] Core nets are analogous to expanded cores for theories of mathematical physics in [4], p. 179 and [5], p. 130.

[17] $A(N^*)$ corresponds to $N_{\mathfrak{g}}$, [4], p. 181, and $A_e(E)$, [5], p. 133.

[18] This appears to be an advantage over the earlier 'expanded core' formalism which suggested that the basic theory element had a structurally unique position.

[19] Kuhn has mentioned this sort of little revolution in the Appendix of [2].

[20] 'Weak reduction' and 'reduction' correspond respectively to 'reduction' and 'strong reduction' in p. 216ff. and [5], Ch. 8, Sec. 9. Stylistic grounds motivate the change in terminology – nothing more profound. Stronger reduction relations might be defined by requiring reductive correspondences between individual components in the theory matrices.

[21] I am indebted to Dr C. U. Moulines on this point.

BIBLIOGRAPHY

[1] Balzer, W. and Sneed, J. D.: (to appear), *Inter-theoretical Structures in Empirical Science*.

[2] Kuhn, T. S.: 1970, *The Structure of Scientific Revolutions* (2nd ed.), Chicago.

[3] Moulines, C. U.: 'A Logical Reconstruction of Simple Equilibrium Thermodynamics', *Erkenntnis* **9**, (1975), 101–130.

[4] Sneed, J. D.: 1971, *The Logical Structure of Mathematical Physics*, D. Reidel, Dordrecht.

[5] Stegmüller, W.: 1973, *Probleme und Resultate der Wissenschaftstheorie und Analytischen Philosophie, Band II, Theorie und Erfahrung: Zweiter Halb-band, Theorien Strukturen und Theoriendynamik*. Springer.

[6] Suppes, P. C.: 1961, 'A Comparison of the Meaning and Uses of Models in Mathematics and the Empirical Sciences', in H. Freudenthal (ed.), *The Concept and the Role of the Model in Mathematics and Natural and Social Sciences*, D. Reidel, Dordrecht, p. 163.

ACCIDENTAL ('NON-SUBSTANTIAL') THEORY CHANGE AND THEORY DISLODGMENT*

1. THE TENSION BETWEEN SYSTEMATIC AND HISTORICAL APPROACHES IN THE PHILOSOPHY OF SCIENCE

The philosophy of science as initiated and developed in this century mainly by empiricists was purely systematic in its orientation. Increasing attention to the history of science and to the psychological and sociological aspects of its practice should have, one might have thus expected, meant a welcome addition to the logic of science. Whoever entertained such hopes was, however, in for a bitter disappointment. In particular, with the appearance of Professor Kuhn's work on scientific revolutions it became dreadfully clear that the results achieved in the different branches did not even yield a *consistent* overall picture of science. The fledgling student of the philosophy of science appeared to be faced with having to choose between two *incompatible paradigms:* the logical or the psychological-historical.

Indeed, the situation was even somewhat more aggravated; for the discussion was being conducted at two different levels with the pendulums swinging in opposite directions at each level. At the more concrete level of the philosophy of science the tendency was more and more toward Kuhn's way of looking at things. At the more abstract level of *general* epistemological investigations things looked entirely different. A number of penetrating thinkers attempted to show that even if Kuhn himself did not so intend it, his conception of natural sciences inevitably leads to a form of subjectivism as well as of irrationalism and of relativism, and thus, to positions which for philosophical reasons are untenable or even absurd. Indeed, Kuhn himself often emphasized that these *supposed* consequences of his ideas must be based on misunderstandings. Since, however, it was apparently not possible to pin down the sources of these misunderstandings, these critics did not believe Kuhn.

Butts and Hintikka (eds.), Historical and Philosophical Dimensions of Logic, Methodology and Philosophy of Science, 269–288.

Thus, young philosophers of science were driven into a sort of intellectual schizophrenia. On one hand they found the Kuhnian approach uncommonly attractive, on the other, if they took Kuhn's critics seriously, they felt forced to regard it as in need of fundamental revision.

It has been my firm conviction for a long time now that this represents a wholly impossible situation, and that it is absolutely imperative to bridge the gap between the historically and the systematically oriented approaches. The attempts in this direction to date appeared to me unsatisfactory for many reasons. As I came across Sneed's book it became abruptly clear that at the bottom of Kuhn's theses lies a *theory* concept which is totally different from the one then current among philosophers of science.

For a long time, indeed, for *too* long a time, metamathematics has furnished the model to which philosophers of science turned and on which they attempted to pattern their investigations of theories. For logicians and metamathematicians it goes without saying that theories are classes of sentences. This interpretation had proven so fruitful for handling all problems in these areas that it was never questioned. Philosophers of science adopted this view of theories as a matter of course, and with it, the tacit assumption that in their discipline, too, the logical reconstruction of theories as classes of sentences would prove fruitful. Today, I no longer believe this assumption to be correct. We will gain a better understanding of scientific theories if we give up this statement view. In this connection I should like to follow Bar-Hillel in referring to the new conception positively as the *structuralistic conception of theories*. I hope this short autobiographical excursus will prove helpful in understanding what follows.

In order to forestall false expectations I should like to make two observations before going any further: (1) the task of a logical reconstruction also includes indicating the *limits* of what can be logically comprehended and explained. In regard to the problems at hand, I am convinced that these limits must be much more narrowly drawn than most 'empiricists' and 'rationalists' believe. (2) Although I will at times be dealing with special phenomena, including those which Kuhn calls "normal science" and "scientific revolutions", detailed analyses are not my primary goal here. First and foremost I want to deal with the objections mentioned above, such as irrationalism, subjectivism and relativism, thus contributing to the clarification of questions belonging to the abstract epistemological level.

I would like to specify more closely which systematic approach will be taken here. I have purposely elected the one which presumably lies the furthest from Kuhn's, and is so dissimilar to it that one can scarcely imagine how the two can be brought into touch, namely, the axiomatic method as developed by P. Suppes. The only thing which I was able to discover common to the approaches of Kuhn and Suppes was that both were the target of the most bitter attacks from philosophical quarters, even if for wholly different reasons. While Kuhn reaped the protests already mentioned, Suppes' procedure drew objections primarily on grounds that it is so abstract and so general that it precludes a discussion of a host of problems central to the philosophy of science. Essentially, these objections culminate in the challenge to specialize Suppes' method in such a way that the epistemological problems in question can be discussed. It appears to me that Sneed has achieved this to a great extent. Each one of the realistic, pragmatic steps taken by Sneed in the process of specializing the Suppes approach constitutes simultaneously a step toward the erection of a new pillar for the bridge leading to the historically oriented philosophy of science.

At the risk of being repetitious I will again state the essential points. First, the idea of a single 'cosmic' application of a physical theory is scrapped in favor of the thesis that each such theory has *several partly overlapping applications*. Second, these intersections lead to the important *differentiation between laws and constraints*. While laws hold in some one, or possibly all applications, constraints establish more or less strong 'cross-connections' between the particular applications. Third, it should be remembered that the *special laws* holding only for certain applications must be differentiated from the basic law which is to be incorporated into the core of a theory. A fourth point appears to me to be of utmost importance, namely, the *theoretical-nontheoretical dichotomy*. I should like to remind you that this distinction is handled quite differently than it was within the framework of empiricism. The scientific language is not divided into a 'fully understandable observational language' and a theoretical language which is 'only partly interpretable' by means of correspondence rules. Instead, the theoretical terms of a theory T are distinguished on the basis of a criterion. The measurement of theoretical functions depends upon a successful application of just this theory T. Thus, one can say that these quantities are T-determinable, and one must henceforth speak of 'T-theoretical quantities', not simply 'theoretical quantities'. It appears to me that only in this way do we find an answer to "Putnam's challenge", namely, to show 'in what way theoretical terms come from theories'. The theoretical terms 'come from the theory' in the sense that their values are measured in a theory-dependent way. This leads to Sneed's problem of theoretical terms whose only known solution to date is the Ramsey-method. The contrast to the traditional way of thinking becomes abundantly clear where two different physical theories T_1 and T_2 are formulated in the same language. One and the same term of this language can then be simultaneously theoretical and non-theoretical; i.e. theoretical in relation to T_1 and non-theoretical in relation to T_2.[1] Later I will consider a fifth point, *the method of paradigmatic examples.*

2. THE STRUCTURALISTIC THEORY CONCEPT: THEORIES, THEIR EMPIRICAL CLAIMS, HOLDING A THEORY

I would now like to examine more closely the structuralistic view of theories. Compared with traditional ideas identifying theories with classes of sentences, this structuralistic conception offers five important advantages: (1) With it a concept corresponding to Kuhn's notion of "*normal science*" can be introduced in an unforced natural way. (2) With it a *concept of progress* can be introduced which also covers the revolutionary cases. (3) The phenomenon of the immunity of theories to 'recalcitrant experience' can be made clear and understandable. (4) It permits an elegant simplification of what Lakatos intended his *theory of research programmes* to achieve. (5) It removes the danger – and this is perhaps the main advantage – of falling into a rationality monism and, thus, into the *rationality rut* of assuming there could be but *one single* source of scientific rationality.

We now turn again to the central question of what may be understood as a *theory.* As you remember Sneed has already distinguished between

(1)   a theory as an entity based on a theory element $\langle K, I \rangle$,

whereby K is a theory core and I is the set of intended applications,

and

(2)   the empirical claims of a theory, which have the form $I \in \mathbb{A}(N^*)$ with N^* being the core-net induced by the theory-net N.

If it is asked whether (1) or (2) applies to *that which a scientist considers a theory*, the concrete answer in the most cases will be neither. The scientist will be thinking of something much less abstract, something with flesh and blood involving people, their convictions and their knowledge. This third concept we will call *holding a theory*.

Perhaps the following analogy to the philosophy of language will help to make this somewhat clearer. Sneed occasionally characterized theories and empirical claims as *products*. Linguistic objects, words and sentences, can also be seen as products, and speech act theory has shown that the extremely important dimension of the performative modi is lost from this point of view. Similarly, theories and empirical claims stand in much the same relation to *acts of holding* a theory as do linguistic products to speech acts.

The concept of holding a theory can be introduced in a broader or in a narrower sense. Further, one can give it a more objective or a more subjective accent. For all these definitions one needs extra-logical concepts such as 'person', 'believes that', 'has supporting evidence for', as well as a variable t ranging over historical times. Instead of formal precision I will attempt here an intuitive gloss.[2] In order to introduce the *weak* objective variant, we will assume a theory T in the earlier sense to be given. The statement that a person p holds a theory T (in the weak sense) at time t (abbreviated $H_w(p, T, t)$) means that there is a net N based on specializations of the basic core K_b of T such that p believes $I \in \mathbb{A}(N^*)$ at t, furthermore, that p has supporting evidence for this proposition, and finally that p believes N to be a *strongest* existing net such that $I \in \mathbb{A}(N^*)$. If one likes, one can also incorporate into this concept the person's belief that using this theory will yield progress. The phrase 'has supporting evidence for' implicitly contains the confirmation problem. Since we are going to disregard this problem here, we will use the abbreviation 'p knows that Y' to mean 'p believes that Y and p has data supporting Y'.

3. 'NORMAL SCIENCE' AND 'SUBJECTIVISM'

Before sketching other variants, I would like to indicate how the concept of holding a theory can be used to explicate the notion of normal science. The idea is this: *if several persons hold the same theory*, they will be said *to belong to the same normal scientific tradition*. This means that the persons in question do indeed use the same theory to construct their hypotheses, but couple it to a variety of different convictions and assumptions. Thus, the theory T remains unchanged, while the empirical claims attached to it may change at any time. I would like to propose calling all those changes not involving the theory itself *accidental theory changes*, since one can draw an illuminating comparison with the ancient substance-accident dichotomy: the basic core $K = \langle M_p, M_{pp}, M, C \rangle$ is the immutable substance underlying change, while the core specializations represent the constantly changing accidents.

But now I am anticipating. To actually arrive at a viable concept of normal science in Kuhn's sense, several important factors must still be considered. To this end I must briefly say something about the concept of *paradigm*. What Wittgenstein had in mind and wanted to illustrate with the example "game" was as follows: Neither an explicit extensional characterization of the predicate G for *game* via a listing *of all* games, nor a precise definition of G by the stipulation of sufficient and necessary conditions for membership in G is possible. Instead, we must limit ourselves to effectively specifying a sub-set G_0 of G, the *list of paradigms* (or *paradigmatic examples*) *of games*. Elements will then be admitted to the difference set $G - G_0$ only if they exhibit a significant number of properties common to most elements of G_0. This formulation underscores the irremediable vagueness adhering to a set determined by paradigmatic examples.

In our case, though, the general concept of a paradigm will only be used for a very special purpose: The set I of intended applications of a theory is not completely specified from the very beginning by means of a list or a strict definition; it is an *open* set for which the theory's creator has stipulated a sub-set I_0 of paradigmatic examples, and which can be changed (through additions and cuts) in the course of working with the theory provided the condition $I_0 \subseteq I$ is met. Thus, for instance, Newton specified the paradigms for the application of his theory by designating examples such as the solar system, certain parts thereof, the tides, pendulum motion, and free falling bodies near the earth's surface

This idea can be utilized for our explication attempt as follows:[3] We expand the present theory concept to a concept of a theory in the strong sense, i.e. to what Sneed called a Kuhn-theory, and that there once existed a person p_0 who at time t_0 successfully applied the core K of the theory, i.e. $I_0 \in A(K)$. We then modify the concept of holding a theory by taking "theory" to mean this strong theory concept and requiring that the person p holding the theory also choose the set I_0 as the set of paradigms for I. This establishes the *historical source* of the theory in its inventor, as well as the *paradigm* concept and the *historical continuity* between all those persons holding the theory. The resulting *concept of holding a theory in the strong sense*, $H_{st}(p, T, t)$, could be taken as an explication of the Kuhnian concept of normal science, at least of its objective variant.

This objective variant can be replaced by a subjective one. The only difference is that *some existential quantifiers and the epistemic operators 'believes that' and 'knows that' are switched around*. Whereas previously we always spoke of the existence of a net N and a set I about which the person p knows or believes something, we now say that p knows of a net N and a set I that $I^0 \subseteq I$ and $I \in A(N^*)$, and that as far as p knows N is a strongest net and I a largest set of its kind. If under 'subjectivism' nothing other is meant than a philosophical temperament which prefers this subjective to the objective variant, this would be a quite viable subjectivism.[4]

4. RATIONALITY AND PROGRESS BRANCHING IN NORMAL SCIENCE

What about the rationality of the normal scientist? In principle this question is easy to answer; indeed, without having to go into the problem of whether, and how, criteria for scientific rationality can be formulated. For whether these criteria are inductivistic, deductivistic or something else, their satisfaction, or violation, pertains only to empirical claims, and thus turns completely on the word 'knows'. In any case the normal scientist, i.e. the scientist holding a given theory, *can* satisfy any of these rationality criteria.

Normal science allows for *two sorts of progress*. One consists in expanding the set of intended applications, the other in further specializations of the basic core K_b.

Their counterparts are the corresponding types of *setbacks* which a normal scientist often experiences; namely, being forced to retract an attempted expansion of the range of application or an attempted core specialization.

I would also like to mention two interesting complications. Wherever we have used a superlative to characterize the concept of holding a theory, it was always with the *indefinite*, not the definite article. The reason lies in the following possibility:

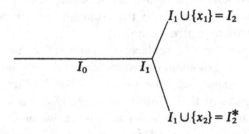

$$I_1 \cup \{x_1\} = I_2$$
$$I_0 \qquad I_1$$
$$I_1 \cup \{x_2\} = I_2^*$$

The branching is intended to indicate that a given core specialization is applicable either to I_2 or to I_2^* but *not* to $I_1 \cup \{x_1\} \cup \{x_2\}$. In such a situation neither the theory, nor experience, nor logical reasoning can help. The scientist must *decide* on the basis of value judgments.

Another kind of branching is also possible. The scientist can at a given time be faced with the choice of either expanding his current range of application I at the expense of additional expansions of his net, or leaving I unchanged in order to gain a further core specialization.

Typical situations of this kind could be called *progress branching in normal science*. In these branches we have located a juncture where value judgments are unavoidable in deciding which way to proceed.

Should someone regard *this* as subjectivism, the *only* correct reply is that *this is a species of subjectivism which we can not evade.*

5. HOLISM OF EMPIRICAL CLAIMS. THE THEORY-LADENNESS OF OBSERVATIONS

In connection with this reconstruction sketch for the concept of normal science, I should like to make a few short remarks about two concepts which appear often in the current literature. The expression *"holism"* can be used in relation to theories as well as empirical hypotheses. In relation to empirical content the holistic standpoint can be formulated as follows (and this is a true statement): *the empirical content of a theory at a certain time is not exhibited by numerous special hypotheses constructed on the basis of this theory, but by one single big empirical claim:* $I_t \in A(N_t^*)$.

Concerning the second item, the so-called *theory-ladenness of observations,* I would like to point out an equivocation which has caused much confusion. It is maintained, for example, that the description of the facts relevant for a theory, itself requires a theory. This is in most cases true. But it is also harmless, and creates no special problem. For the theory required is, of course, not the same as that *for which* the facts are being described, but a more elementary, underlying theory. Some authors want, though, to maintain something much stronger than this, namely, that the facts for a theory T are determined *by this theory itself.* This appears to be the meaning of such phrases as 'theories define their own facts'. Yet even this stronger version of the thesis is not only intelligible, but correct *if confined to T-theoretical terms.* And in this case it creates a serious problem, namely, the problem of theoretical terms as formulated by Sneed for which only the Ramsey-solution is currently known.[5]

6. THEORY DISLODGMENT WITHOUT FALSIFICATION. THE THREEFOLD IMMUNITY OF THEORIES. THEORY CHOICE AND RATIONALITY

The concept of holding a theory, which served to define the notion of normal science, or at least an important aspect of this notion, was only the first step toward de-irrationalizing the current image of the Kuhnian conception of science. Now what about scientific revolutions? Can the logician also contribute to a better understanding of *this* phenomenon? Here I should like to begin with a confession so that you will not be too terribly disappointed with the following remarks: The logician can actually accomplish far less in this case than in the case of the phenomenon which Kuhn called normal science. This is not because scientific revolutions are in fact thoroughly irrational processes, but simply because many aspects of these phenomena lie outside the competence of the logician.

First we will try to characterize that aspect of the phenomenon described by Kuhn which again shocked many readers and led to charges of irrationalism, subjectivism and this time relativism as well. All empiricist philosophers, and the modern rationalists too, agreed until quite recently that a theory which founders on experience must be discarded.

As opposed to this, Kuhn's thesis is that even a theory plagued with ever so many anomalies is not discarded because it has foundered on experience. Instead, it is jettisoned only when another theory is available to take its place. This prima facie curious phenomenon may be called '*theory dislodgment by a superseding theory*', or briefly '*theory dislodgment*'. I will attempt to show that in this case, too, the structuralistic view of theories enables us first, to gain a basic logical understanding of the situation; second, to produce a plan for a viable concept of scientific progress in revolutionary changes; and thus, third, to deliver something like a logical test for the correctness of Kuhn's thesis. Concerning the first point we must focus our attention upon a certain particular aspect of theories, namely, their *steadfastness in the face of 'recalcitrant data'*. This is expressly emphasized by Kuhn and felt by many to be especially shocking.

Such an immunity does actually exist and, indeed, in three different respects. The by far most important of these, the *first kind of immunity*, arises when one considers the relationship between the *core* of a theory and the specializations of this core. As you will remember, the empirical claims of a theory with the basic core K_b have the form: $I_t \in \mathbb{A}(N_t^*)$ (whereby 't' is a historical time index). As an empirical claim this 'central empirical statement' of a theory can be refuted by experience.

This refutation does not, however, directly effect the theory itself, for the falsification of the empirical claim only proves that *certain* specializations of the core are not suitable. Indeed, *the same holds for every finite number of such unsuccessful attempts.* We can, by paralleling Popper's argument proving the non-verifiability of strict universal quantifications, obtain the following proof for the unrefutability of a theory on the basis of a finite number of refuted empirical claims: *since the number of possible specializations of a theory-element is potentially infinite, no number, be it ever so large, of unsuccessful attempts to specialize a given theory-element can be considered conclusive proof that a successful specialization of this element is impossible.* Thus, we are not forced to give up the theory; there might just be a still undiscovered specialization which would prove successful when discovered.

If we place ourselves in the 'normal scientist's' situation, i.e. in the situation of one who already holds a theory, we will realize that this scientist is always working with theories whose cores have in the past repeatedly served well. When, therefore, a member of the community is not successful in working with the core, it is natural and understandable that he, and not the theory, be blamed.

Many philosophers have thought that the description of the phenomenon of theory dislodgment indicated a rationality gap. In particular, Kuhn's thesis that the decision to scrap a theory *is always simultaneously a decision in favor of a new theory* was thought to imply something irrational and, thus, logically incomprehensible.[6] To support this, their second charge of irrationalism, they argued that there must be something like a *critical level* at which a theory must be rejected regardless of whether a new one is available or not. Here again we clearly discern the influence of the statement view, namely in the attempt to put scientific theories in the same category with *statistical* hypotheses, if not deterministic laws. As the brief logical analysis has shown, there is no critical level at which a theory must be discarded. Stipulating such a level would be a purely arbitrary act.

Concerning the second half of Kuhn's thesis we need only add a psychological truism to the immunity already established in order to comprehend this situation, namely, the elementary insight that people do not throw away tools which have served well as long as they possess no better substitutes. Or to take a more drastic example which better

depicts the situation of a crisis-ridden scientific theory: Would someone who was freezing not seek shelter in a hut simply because it was
awfully ramshackle? Were he not to, this would mean he prefers sure
death to mere danger.

That the philosopher has overstepped his competence here by resorting to *empirical generalizations* is a charge I could not accept. The
situation seems to me similar to that of the ordinary-language
philosopher. He does not base his analyses on statistical surveys
concerning the use of language, but on his own linguistic competence.
In respect of psychological truisms like the one just mentioned, we
have at least the same degree of competence. As a *human being* I am
competent to evaluate certain reactions as *typically human* without
having to resort to generalizations, which properly ought to be called
'hypothetical'.

These two forms of understanding, based on the linguistic competence of the native speaker and on the competence of our judgments
concerning spontaneous human reactions respectively, could be called
elementary hermeneutic understanding. If we accept this choice of
terminology, we can say that in order to gain an accurate grasp of the
phenomenon of theory dislodgment, logical and hermeneutic understanding must work together.

Such cases must, however, be carefully distinguished from those
situations in which a negative decision is made in relation to certain
parts of the range of application of a theory. In order to see this clearly
we must take a look at the *second kind of theory immunity*. To this end
we recall once again the relationship between I_0 and I. I_0 is the
explicitly given extensional sub-set of I consisting of the paradigmatic
examples. This sub-set can never be changed. With the exception of
this minimal requirement, I is an open set. Should the scientific
community in trying to apply the theory to some $a \in I - I_0$ experience
fundamental difficulties, stretching perhaps over generations of research, and thus, conclude that this application is not possible, it need
not, *contrary to falsificationism*, hold the theory responsible. Instead, it
can decide to deny a's membership in the theory's range of application.

Here we come in sharp conflict with the demands of the 'critical
rationalists'. For according to their notion of critical attitude, a scientist

should make his theory *as sensitive as possible* to potential refutation. In our present case that would mean stipulating sharp criteria for membership in *I* and, consequently, rejecting the theory. I would like to counter this demand with the following observation: it appears that no physicist has ever been willing to assume the risk of falsification involved in stipulating sufficient and necessary conditions for membership in *I*. When optical phenomena could not be explained with the help of Newton's particle mechanics, as he hoped they could, his work was not pronounced invalid; instead it was concluded in accord with Maxwell's conception that light did not consist of particles. As I have not approached science via a preconceived over-all conception of scientific rationality I am unable to perceive anything irrational in such a decision.

But what happens when the fundamental law of the theory fails in I_0? Here we meet the *third* kind of immunity of theories. It is a consequence of two particular features of physical theories: first, the occurrence of theoretical terms in the core, and second, what I have called the holism of empirical claims. Instead of analyzing the general case we will illustrate with a simple example. Assume we agree to take Newton's second law as the only fundamental law of his theory. Then accepting this law means nothing more than being committed to promise that suitable force and mass functions satisfying this law exist in all intended applications, functions which take special forms in certain applications and are connected across these applications by certain constraints. It is the near vacuousness of this wide-ranging promise which precludes its empirical refutation.

7. THEORY HOLISM AND 'PROPAGANDA'. THE ROLE OF VALUE JUDGMENTS

In view of these three forms of irrefutability it is understandable that even a theory caught in a crisis will almost always be retained until a promising new theory is constructed. Here another viable form of *holism* takes its place along side the holism of empirical claims. This *'theory holism'* could be briefly formulated as follows: the decision to accept a theory is always *an all-or-nothing-decision,* and it *cannot be replaced by any rules nor dictated by a so-called experimentum crucis.*[7]

8. OVERCOMING THE RELATIVISM CHARGE.
PROGRESSIVE REVOLUTIONS

Following this attempt to de-irrationalize and de-subjectivize the current picture of Kuhn's conception of science, I must now take up the *charge of relativism.* It seems to contain a grain of truth.

For the sake of illustration I will briefly formulate the relativism charge against Kuhn as sharply as possible: 'An actual change of theories is described by Kuhn in sociological-psychological language. In this way one gets the impression of a complete parallel with religious and political power struggles. Assume further that to this picture be added the thesis of the incommensurability of the old and the new theories. All this adds up to *relativism.*'

We want, now, to illustrate this objection with the following possible-world picture: Let T_1 and T_2 be two theories designed to solve the same kind of problems but having different theoretical terms. Initially theory T_1 prevails in the possible world w_1, but is subsequently dislodged by theory T_2 after having run into a crisis. Exactly the opposite transpires in possible world w_2. Here T_2 prevails initially, is beset by crisis and subsequently dislodged by T_1. One can well imagine this happening given different psychological and sociological conditions suitable to each case. In both worlds the proponents of the new theory are convinced they have brought progress. Thus, left alone with the psychological-sociological progress criterion, we must admit *that in both worlds there has been, by definition, progress.* This appears fully unacceptable to us. 'Revolutionary progress' *cannot designate a symmetrical relation.*

It is important not to misunderstand my point. I do not deny that two such events could *happen,* nor that in both cases those involved *believe* progress has been served. My thesis is simply that in at least one of these two worlds the agents of the *alleged* progress must be mistaken; real progress can only have taken place in one of the two worlds.

Thus we have located the heart of the problem. The 'missing link' by Kuhn is not a 'critical rejection level'[8]; it is the *introduction of an adequate concept of scientific progress for the case of theory dislodgment*. Only in this way, it appears to me, can we avoid the Scylla of teleological metaphysics and the Charybdis of relativism.

Today I believe that Kuhn himself wanted to point out this difficulty. The last pages of his book can be understood as a challenge to the systematic philosophy of science to come up with a viable concept of scientific progress not infected with a teleological metaphysics. In my opinion Sneed's reduction concept offers the best start in this direction to date, if not yet the final solution. It also appears to me that the late Professor Lakatos had something similar in mind with his concept of sophisticated falsification. Sneed's reduction concept is especially interesting because it permits *a comparison of theories with fully heterogeneous conceptual apparatus.*

In cases of radical theory change a complete reduction is presumably not possible. Here one must be satisfied with the notion of the *approximative imbedding* of one theory in another. Sneed's reduction concept must be correspondingly expanded and liberalized. Concerning this extra-ordinarily important approximation problem there is, as far as I know, only one single interesting preliminary work. I mean a book from Günther Ludwig.[9] Ludwig works with '*blurred*' functions and relations, indeed, even with '*blurred*' objects. This idea may be successfully incorporated into the Sneedian conceptual apparatus as my collaborator and former student Moulines has shown. In this way it becomes possible to work with *blurred* models and possible models, even with *blurred* partial possible models.[10] Nevertheless I must admit that this is not yet a fully developed theory.

I must append a qualifying remark to this discussion of theory dislodgment. I have for simplicity's sake considered only the case of '*radical*' theory change; i.e., the case where the basic core K_b is replaced by another. The construction of theory nets makes it possible to give an analogous account of a 'mini-revolution' such as Kuhn calls attention to in the appendix to the 2nd edition of his book. We have this kind of 'small revolution' when the 'pyramid point' K_b remains unaltered but a specialization *at a relatively high place in the pyramid* is replaced by another. On the other hand it is also possible that during the course of the development of a theory not only the 'pyramid point' K_b but also certain specializations remain unaltered.

9. Is progress branching possible in revolutionary theory dislodgments? the 'evolutionary tree'

Is there such a thing as revolutionary progress branching? Concerning this question I would like to present a provoking thesis. According to teleological conceptions of scientific progress there surely can be no such thing. But the concepts of reduction and approximative imbedding were intended primarily to serve in formulating an *immanent* progress criterion which dispenses with all such 'metaphysical conceptions'. Assume now for the sake of argument that such progress concepts are available. It is then quite conceivable that we run across situations of the following kind:

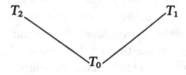

This diagram is to be interpreted as follows: a theory T_0 can be dislodged by either T_1 or T_2, whereby T_0 is either *reducible to or approximatively imbeddable in both*. T_1 and T_2 are nevertheless neither equivalent, nor is either one reducible to the other because although both T_1 and T_2 explain the phenomena explained by T_0, the totality of phenomena which the one accounts for only *partially overlaps* those explained by the other. In perfect analogy to the case of normal scientific progress branching, we have here a juncture at which *ultimate*, not provisional, *value judgments* must decide which route to take, or whether both such paths should be pursued.

Add this possibility to the two forms of normal scientific progress branching already described and we see that from a logical point of view nothing can be urged against the picture of a branching 'evolutionary tree' which shocked so many. Nevertheless, many philosophers will object that the mere belief in the *conceivability* of such a situation implies a scientific relativism. To this I would reply: if by definition any interpretation of scientific progress not logically producing linearity is to be called 'relativism', indeed, even when an adequate progress concept is available, *then this is presumably a form of relativism which we must swallow.* My initial position must then be re-formulated. It is true that I tried to de-irrationalize and de-subjectivize the picture of Kuhn's conception of theory change shared by philosophers of quite different persuasions. On the other hand, though, I have not only not furnished the means for overcoming a certain variant of what some call *relativism,* but I have attempted to show that this form of relativism is *defensible.* If, however, it is asked whether, and how often, such branching has actually occurred, the logician, *qua logician,* is unable to say anything and must pass the question along to his colleagues in other fields. But it must not be overlooked that no matter what the answer, there would still remain a problem to be solved. For even if progress branching has occurred, it was presumably quite rare. But why? I know of no general answer. We are faced with a somewhat paradoxical situation: the prima facie shocking idea of progress producing a branching evolutionary tree proves under closer analysis to be epistemically harmless. The *real* problem here is to explain *why such branching is much more rare than one would expect.* I have only the vague idea that an adequate answer will involve peculiarities of human nature as well as internal and external factors.[11]

Let me make a concluding remark. I cannot hope that the material presented by Sneed and myself will suffice to convince you. On the one hand you may have the impression of having been bombarded with too many novelties. On the other, we have for the time being only a meager basis, namely, the theories of mathematical physics. Whatever a logician may have to say about a topic like 'theory change', it is sure to be dry as bones compared with the fascinating vividness and colorful richness of Kuhn's writings.

Since the logic of science is still in its infancy very much remains to be done before the bridge between it and the history of science can be completed. Future success depends, however, not only on our efforts and skills, but on something else too. In those fields of knowledge dealing with human affairs one can observe in recent times an unfortunate trend. The representatives of various schools of thought do not even listen to one another any more.[12] This trend has begun to catch hold in the philosophy of science too, and seems to gain momentum the more opinions diverge.

This need not be, since it has not always been so. Ancient philosophers differed in their opinions no less than philosophers today. But no matter how vigorously they attacked each other they never refused to talk it out. The future situation in the philosophy of science will depend to a great extent on whether we succeed in regaining this virtue of the ancient Greeks, namely, of listening to each other.

NOTES

* A version of this paper was published in advance in *Erkenntnis* **10** (1976), 147–178.
[1] The concept 'pressure' offers an example if we take mechanics as T_1 and phenomenological thermodynamics as T_2.
[2] For formal definitions cf. Sneed, [10], p. 266, and Stegmüller, [11], p. 194.
[3] For technical details cf. Sneed, [10], p. 294, and Stegmüller, [11], pp. 221ff. The deviation of the latter from Sneed's definition is due to the attempt to eliminate the 'platonistic' character of the set *I*. This was done by way of an auxiliary definition (D15, p. 221) in which the applications of a theory *T* accepted by a person at a given time were defined.
[4] For a more exact formulation of this *subjective variant* cf. Stegmüller, [12].
[5] Cf. Stegmüller, [11], holistic thesis (III), p. 272 and its discussion p. 276.
[6] This view was advocated, e.g., by J. Watkins in [13], as well as originally by I. Lakatos in [5].
[7] Cf. Sneed, [10], pp. 90ff. and Stegmüller, [11], pp. 271ff. Three further kinds of 'holism' may be distinguished from the two mentioned here ('holism of empirical claims' and 'holism of empirical theories'). Sometime the thesis of the '*theory-ladenness of observations*', according to which 'a theory defines its own facts', is included as a part of this thesis (cf. Stegmüller, [11], holistic thesis (III) pp. 272ff.). Besides these, there is also what could be called the '*holism of refutation and confirmation*'. According to it only an entire scientific system can be confronted with 'experience' and supported or falsified by it. A detailed examination of this form of holism is a task for confirmation and test theory. But in any case, one can extract a certain concession to this conception from the 'holism of empirical claims' described in Section I. Finally, there also appears to be something like a '*methodological holism*'. According to it one must answer all questions in the philosophy of science simultaneously. Theories, for example, cannot be investigated in isolation, but only in the context of the entire theory hierarchy. Or the problem of scientific explanation must be dealt with together with the problems of concept and theory construction, and these in turn with those of confirmation and testing. This fifth

form of holism is unacceptable for the simple reason that it makes superhuman demands on philosophers.
[8] Were there one, the phenomenon of theory dislodgment as such would have to be regarded either as an irrational process or as an incomplete description of a rational process. The present location of the 'missing link' is such that it need neither contest the phenomenon of theory dislodgment nor regard it as an incomplete description.
[9] G. Ludwig, [7], pp. 71ff.
[10] This theory, which uses quite strong topological concepts, cannot be sketched here. The starting point is the concept of *uniform filters* as introduced by G. Ludwig, op. cit., p. 76f.
[11] Internal factors would, for example, be pre-systematic intuitive considerations of analogy or simplicity. The external factors would include, among other things, available technology and the dominating Weltanschauung. Concerning the latter, cf. also K. Hübner, 'Zur Frage des Relativismus und des Fortschritts in den Wissenschaften. Imre Lakatos zum Gedächtnis', *Journal for the Philosophy of Science*, 5, (1974) 285–303.
[12] An impressive account of this deplorable situation within social philosophy has been given by Kurt v. Fritz in his booklet: *The Relevance of Ancient Social and Political Philosophy for our Times. A short Introduction to the Problem.* Berlin-New York, 1974.

BIBLIOGRAPHY

[1] Kuhn, T. S.: 1970, *The Structure of Scientific Revolutions* (2nd ed.), University of Chicago Press, Chicago.
[2] Kuhn, T. S.: 1970, 'Logic of Discovery or Psychology of Research?', in I. Lakatos and A. Musgrave (eds.), *Criticism and the Growth of Knowledge*, Cambridge University Press, Cambridge, pp. 1–23.
[3] Kuhn, T. S.: 1970, 'Reflections on My Critics', in I. Lakatos and A. Musgrave (eds.), *Criticism and the Growth of Knowledge*, pp. 231–278.
[4] Kuhn, T. S.: 1973, 'Objectivity, Value-Judgment, and Theory Choice', The Franklin J. Machette Lecture, Furman University.
[5] Lakatos, I.: 1970, 'Falsification and the Methodology of Scientific Research Programmes', in I. Lakatos and A. Musgrave (eds.), *Criticism and the Growth of Knowledge*, pp. 91–195.
[6] Lakatos, I.: 1972, 'History of Science and Its Rational Reconstruction', *Boston Studies in the Philosophy of Science*, Vol. VIII, D. Reidel Publishing Company, Holland, pp. 91–136.
[7] Ludwig, G.: 1970, *Deutung des Begriffs "physikalische Theorie" und axiomatische Grundlegung der Hilbertraumstruktur der Quantenmechanik durch Hauptsätze des Messens*, Springer-Verlag, Berlin-Heidelberg.
[8] Moulines, C.-U.: 1975, *Zur Logischen Rekonstruktion der Thermodynamik*, Diss., Universität München.
[9] Moulines, C.-U.: 1975, 'A Logical Reconstruction of Simple Equilibrium Thermodynamics', *Erkenntnis* 9, 101–130.
[10] Sneed, J. D.: 1971, *The Logical Structure of Mathematical Physics*, D. Reidel Publishing Company, Dordrecht.

[11] Stegmüller, W.: 1973, *Probleme und Resultate der Wissenschaftstheorie und Analytischen Philosophie, Band II, Theorie und Erfahrung: Zweiter Halbband, Theorienstrukturen und Theoriendynamik*, Springer-Verlag, Berlin-Heidelberg. English Translation Springer-Verlag, New York, 1976.

[12] Stegmüller, W.: 1975, 'Structures and Dynamics of Theories. Some Reflections on J. D. Sneed and T. S. Kuhn', *Erkenntnis* **9**, 75–100.

[13] Watkins, J.: 1970, 'Against 'Normal Science',' in I. Lakatos and A. Musgrave (eds.), *Criticism and the Growth of Knowledge*, pp. 25–37.

THOMAS S. KUHN

THEORY-CHANGE AS STRUCTURE-CHANGE: COMMENTS ON THE SNEED FORMALISM*

1. INTRODUCTION

It is now more than a year and a half since Professor Stegmüller kindly sent me a copy of his *Theorie und Erfahrung* (Stegmüller, 1973), thus drawing my attention for the first time to the existence of Dr. Sneed's new formalism and its likely relevance to my own work. At that time set theory was to me an unknown and altogether forbidding language, but I was quickly persuaded that I must somehow find time to acquire it. Even now I cannot claim entire success: I shall here sometimes refer to, but never attempt to speak, set theory. Nevertheless, I have learned enough to embrace with enthusiasm the two major conclusions of Stegmüller's book. First, though still at an early stage of its development, the new formalism makes important new territory accessible to analytic philosophy of science. Second, though sketched with a pen I can still scarcely hold, preliminary charts of the new terrain display remarkable resemblance to a map I had previously sketched from scattered travellers' reports brought back by itinerant historians of science.

The resemblance is firmly underscored in the closing chapter of Sneed's book (Sneed, 1971, esp. pp. 288–307); its detailed elaboration is a primary contribution of Stegmüller's. That the rapprochement both see is genuine should be sufficiently indicated by the fact that Stegmüller, approaching my work through Sneed's, has understood it better than any other philosopher who has made more than passing reference to it. From these developments I take great encouragement. Whatever its limitations (I take them to be severe), formal representation provides a primary technique for exploring and clarifying ideas. But traditional formalisms, whether set-theoretical or propositional, have made no contact whatsoever with mine. Dr Sneed's formalism does, and at a few especially strategic points. Though neither he, nor

Butts and Hintikka (eds.), Historical and Philosophical Dimensions of Logic, Methodology and Philosophy of Science,

Stegmüller, nor I suppose that it can solve all the outstanding problems in philosophy of science, we are united in regarding it as an important tool, thoroughly worth much additional development.

Just because the new formalism does illuminate some of my own characteristic heresies, my evaluation of it is unlikely to be free from bias. But I shall not pause merely to deplore the inevitable. Instead, I turn to my subject proper, beginning with a cursory sketch of some aspects of the new formalism that seem to me particularly appealing. Premising them, I shall next explore two aspects of the Sneed-Stegmüller position that, in their present form, seem to me significantly incomplete. Finally, I shall examine one central difficulty that will not be resolved within the formalism but presumably requires resort to philosophy of language. Before turning to that program, however, let me avoid misunderstanding by indicating an area in which this paper makes no claims at all. What has excited me about the Sneed formalism is the issues it makes it possible to explore with precision, not the particular apparatus developed for that purpose. About such questions as whether or not those achievements demand the use of set and model theory, I have no basis for opinions. Or rather, I have a basis for only one: those who think set theory an illegitimate tool for analysing the logical structure of scientific theories are now challenged to produce similar results in another way.

2. Appraising the Formalism

What has struck me from the start about the Sneed formalism is that even its elementary structural form captures significant features of scientific theory and practice notably absent from the earlier formalisms known to me. That is perhaps not surprising, for Sneed has repeatedly inquired, while preparing his book, how theories are presented to students of science and then used by them (e.g. Sneed, 1971, pp. 3f., 28, 33, 110–114). One result of this procedure is the elimination of artificialities that have in the past often made philosophical formalisms seem irrelevant to both practitioners and historians of science. The one physicist with whom I have to date discussed Sneed's views has been fascinated by them. As an historian, I shall myself mention below one way in which the formalism has already begun to influence my work. Though even guesses about the future are premature, I shall risk one. If only simpler and more palatable ways of representing the essentials of Sneed's position can be found, philosophers, practitioners, and historians of science may, for the first time in years, find fruitful channels for interdisciplinary communication.

To make this global claim more concrete consider the three classes of models required by Sneed's presentation. The second, his potential partial models or M_{pp}'s are (or include) the entities to which a given theory might be applied by virtue of their description in the non-theoretical vocabulary of the theory. The third, his models or M's derive from the subset of the M_{pp}'s to which, after suitable theoretical extension, the laws of the theory actually do apply. Both find obvious parallels in traditional formal treatments. But Sneed's partial models, his M_p's, do not. They are the set of models obtained by adding theoretical functions to all the suitable members of M_{pp}, thus completing or extending them prior to the application of the theory's fundamental laws. It is in part, by giving them a central place in the reconstruction of theory that Sneed adds significant verisimilitude to the structures that result.

Lacking time for an extended argument, I shall be content here with three assertions. First, teaching a student to make the transition from partial potential models to partial models is a large part of what scientific, or at least physics, education is about. That is what student laboratories and the problems at the ends of chapters of textbooks are for. The familiar student who can solve problems which are stated in equations but cannot produce equations for problems exhibited in laboratory or stated in words has not begun to acquire this essential talent. Second, almost a corollary, the creative imagination required to find an M_p corresponding to a non-standard M_{pp} (say a vibrating membrane or string before these were normal applications of Newtonian mechanics) is among the criteria by which great scientists may sometimes be distinguished from mediocre.[1] Third, failure to pay attention to the manner in which this task is done has for years disguised the nature of the problem presented by the meaning of theoretical terms.

Except in the case of fully mathematized theories, neither Stegmüller nor Sneed have much to say about how M_{pp}'s are, in fact, extended to M_p's. But the view Sneed develops with precision for his special case is strikingly like the one I had earlier articulated vaguely for the general one, and the two may henceforth fruitfully interact, a point to which I shall return. In both cases the process of extension depends upon assuming that the theory has been correctly deployed in one or more previous applications and on then using those applications as guides to the specification of theoretical functions or concepts when transforming a new M_{pp} to an M_p.[2] For fully mathematical theories that guidance is supplied by what Sneed calls constraints, law-like restrictions that limit the structure of pairs or sets of partial models rather than of individual ones. (The values assumed by theoretical functions in one application must, for example, be compatible with those assumed in others.) Together with the correlated notion of applications, that of constraints constitutes what I take to be the central conceptual innovation of Sneed's formalism, and another especially striking one follows from it. For him as for me, the adequate specification of a theory must include specification of some set of exemplary applications. Stegmüller's sub-section, 'Was ist ein Paradigma?', is a splendid elaboration of this point (Stegmüller, 1973, pp. 195–207).

So far I have mentioned aspects of Sneed's formalism which cohere particularly closely with views I have developed in other places. Shortly I shall return to some others of the same sort. But I am not sure that closely associating our views will prove a favor to him, and there are other reasons to take his seriously. Let me mention just a few closely related ones before returning to my main theme.

Roughly speaking, Sneed represents a theory as a set of distinct applications. In the case of classical particle mechanics, these might be the problems of planetary motion, of pendula, of free fall, of levers and balances, and so on. (Need I emphasize that learning a theory is learning successive applications in some appropriate order and that using it is designing still others?) Considered individually, each application might be reconstructed by a standard axiom system in a predicate calculus (thus raising the standard problem of theoretical terms). But the individual axiom systems would then ordinarily be somewhat different from each other. (Compare Kuhn, 1970, pp. 187–191.) What in Sneed's view supplies their unity, enables a sufficient set collectively to determine a theory, is partly the basic law or laws which all share (say, Newton's second law of motion) and partly the set of constraints which bind the applications together in pairs or at least in connected chains.

With such a set-theoretic structure, individual applications play a double role, one previously familiar at a pre-theoretic level from discussions of reduction sentences. Taken singly, individual applications, like individual reduction sentences, are vacuous, either because their theoretical terms are uninterpretable or because the interpretation they permit is circular. But when applications are tied together by constraints, as reduction sentences are tied together by the recurrence of a theoretical term, they prove capable simultaneously of specifying, on the one hand, the manner in which theoretical concepts or terms must be applied and, on the other, some empirical content of the theory itself. Introduced, like reduction sentences, to solve the problem of theoretical terms, constraints prove also, again like reduction sentences, to be a vehicle for empirical content.[3]

Numerous interesting consequences follow, of which I shall here mention three. Since the discovery that theoretical terms could not readily be eliminated by strict definition, one has been puzzled how to distinguish the conventional from the empirical elements within the process by which they are introduced. The Sneed formalism clarifies the puzzle by giving it additional structure. If a theory, like Newtonian mechanics, had only a single application (for example, the determination of mass ratios for two bodies connected by a spring), then the specification of the theoretical functions it supplies would be literally circular and the application correspondingly vacuous. But, from Sneed's viewpoint, no single application yet constitutes a theory, and, when several applications are conjoined, the potential circularity ceases to be vacuous because distributed by constraints over the whole set of applications. As a result, certain other, sometimes nagging problems change their form or disappear. Within the Sneed formalism, there is no temptation to ask the, to physicists, artificial question whether mass or, alternatively, force should be treated as a primitive in terms of which the other should be defined. Both, for Sneed, are theoretical and in most respects on a par, because neither can be learned or given meaning except within the theory, some applications of which must be presupposed. Finally, perhaps of greatest long-range importance, is the new form taken by Ramsey sentences within Dr Sneed's formalism. Just because constraints as well as laws take on empirical consequences, there are important new things to be said about the function and the eliminability of theoretical terms. (On these subjects see: Sneed, 1971, pp. 31–37, 48–51, 65–86, 117–138, 150–151; Stegmüller, 1973, pp. 45–103.)

These and other aspects of the Sneed formalism deserve and will presumably receive much additional attention, but for me their importance is dwarfed by another, with which this section of my paper concludes. To a far greater extent and also far more naturally than any previous mode of formalization, Sneed's lends itself to the reconstruction of theory dynamics, the process by which theories change and grow. Particularly striking to me, of course, is that its manner of doing so appears to demand the existence of (at least) two quite distinct sorts of alteration over time. In the first, what Sneed calls a theory-core remains fixed, as do at least some of a theory's exemplary applications. Progress then occurs either by discovering new applications which can be identified extensionally as members of the set of intended applications, I, or else by constructing a new theory-core-net (a new set of expansions of the core in Sneed's older vocabulary) which more precisely specify the conditions for membership in I.[4] Both Stegmüller and Sneed emphasize (Sneed, 1971, pp. 284–288; Stegmüller, 1973, pp. 219–231) that changes of this sort correspond to much of the theoretical part of what I have elsewhere called normal science, and I entirely accept their identification. Since by its nature a theory-core is virtually immune to direct falsification, Sneed also suggests and Stegmüller elaborates the possibility that at least some cases of change of core correspond to what I have called scientific revolutions. (Sneed, 1971, pp. 296–306; Stegmüller, 1973, 231–247.)

Much of the rest of this paper is devoted to identifying difficulties with that second identification. Though the Sneed formalism does permit the existence of revolutions, it currently does virtually nothing to clarify the nature of revolutionary change. I see, however, no reason why it cannot be made to do so, and I mean here to be making a contribution towards that end. Even in its absence, furthermore, both my historical and my more philosophical work are illuminated by the attempt to view revolutions as changes of core. In particular, I find that much of my still unpublished research concerning both the genesis of the quantum theory and its transformation during the years 1925–26 discloses changes that can be well represented as juxtapositions of elements from a traditional core with others drawn from one of its recent expansions.[5] That way of regarding revolutions seems to me especially promising because it may shortly permit me for the first time to say something worthwhile about the continuities which endure through them.[6] Work must be done first, however. I shall now begin to suggest what some of it may be.

3. Two problems of demarcation

I have already suggested that the central novelty of Sneed's approach is probably his concept of constraints. Let me now add that it might usefully be awarded a position even more fundamental than the one he attributes to it. In developing his formalism, Sneed begins by selecting a theory, like classical particle mechanics, for which, he emphasizes, strict identity criteria must be presupposed (Sneed, 1971, p. 35; Stegmüller, 1973, p. 50). Examining that theory, he next distinguishes between the non-theoretical and the theoretical functions it deploys, the latter being those which cannot be specified, in *any* of that theory's applications, without resort to the theory's fundamental laws. Finally, in a third step, constraints are introduced to permit the specification of theoretical functions. That third step seems to me just right. But I am far less confident about the two it presupposes, and I therefore wonder about the possibility of inverting the order of their introduction. Could one not, that is, introduce applications and constraints between them as primitive notions, allowing subsequent investigation to reveal the extent to which criteria for theory-identity and for a theoretical/non-theoretical distinction would follow?

Consider, for example, the classical formulations of mechanics and electromagnetic theory. Most applications of either theory can be carried through without recourse to the other, a sufficient reason for describing them as two theories rather than one. But the two have never been absolutely distinct. Both entered together, and thus constrained each other, in such applications as ether mechanics, stellar aberration, the electron theory of metals, x-rays, or the photoelectric effect. In such applications, furthermore, neither theory was ordinarily conceived as a mere tool to be presupposed while creatively manipulating the other. Instead, the two were deployed together, almost as a single theory of which most other applications were either purely mechanical, on the one hand, or purely electromagnetic, on the other.[7]

I think nothing of importance is lost by recognizing that what we ordinarily refer to as distinct theories do overlap in occasional important applications. But that opinion depends upon my being prepared simultaneously to surrender any criterion quite so strict as Sneed's for distinguishing theoretical from non-theoretical functions and concepts. What is involved can be illustrated by considering his discussion of classical particle mechanics. Because they can be learned only when some applications of that theory are presupposed, the mass and force functions are declared theoretical with respect to particle mechanics, and they are thus contrasted with the variables space and time, acquired independently of that theory. Something about that result seems to me profoundly right, but I am troubled that the argument appears to depend essentially on conceiving statics, the science of mechanical equilibria, as unproblematically a part of the more general theory that treats of matter in motion. Textbooks of advanced mechanics lend plausibility to that identification of the theory, but both history and elementary pedagogy suggest that statics might instead be considered a separate theory, the acquisition of which is prerequisite to that of dynamics, just as the acquisition of geometry is prerequisite to that of statics. If, however, mechanics were split in that way, then the force function would be theoretical only with respect to statics, from which it would enter dynamics with the aid of constraints. Newton's second law would be required only to permit the specification of mass, not of force.[8]

My point is not that this way of subdividing mechanics is right, Sneed's wrong. Rather I am suggesting that what is illuminating about his argument may be independent of a choice between the two. My intuition of what is is to be theoretical would be satisfied by the suggestion that a function or concept is theoretical with respect to a given application if constraints are required to introduce it there. That a function like force may also seem theoretical relative to an entire theory would then be explained by its manner of entry into *most* of that theory's applications. A given function or concept might then be theoretical in some applications of a theory, non-theoretical in others, a result that does not seem to me troublesome. What it may seem to threaten was, in fact, surrendered long ago, with the abandonment of hope for a neutral observation language.

To this point I have been suggesting that much of what is most valuable in Sneed's approach can be preserved without solving a problem of demarcation raised by his present way of introducing his formalism. But other significant uses of the formalism presuppose distinctions of another sort, and the criteria relevant to them appear to require much additional specification. In discussing the development of a theory over time, both Sneed and Stegmüller make repeated reference to the difference between a theory-core and an expanded-theory-core. The first supplies the basic mathematical structure of the theory – Newton's second law in the case of classical particle mechanics – together with the constraints that govern all the theory's applications. An expanded core contains, in addition, some special laws required for special applications – for example Hooke's law of elasticity – and it may also contain special constraints that apply only when those laws are invoked. Two men who subscribe to different cores *ipso facto* hold different theories. If, however, they share belief in a core and in certain of its exemplary applications, they are adherents of the same theory even though their beliefs about its permissible expansions differ widely. The same criteria for subscribing to one and the same theory apply to a single individual at different times. (Sneed, 1971, pp. 171–184, 266f., 292f.; Stegmüller, 1973, pp. 120–134, 189–195.)

A core, in short, is a structure that cannot, unlike an expanded core, be abandoned without abandoning the corresponding theory. Since a theory's applications, excepting perhaps those which originated with it, depend on specially designed expansions, the failure of an empirical claim made for a theory can infirm only the expansion, not the core, and thus not the theory itself. The manner in which Sneed and Stegmüller apply this insight to the explication of my views should be obvious. Also apparent, I take it, are their reasons for suggesting that at least some changes of core correspond to the episodes I have labelled scientific revolutions. As already indicated, I hope and am inclined to believe that claims of this sort can be made out, but in their present form they have an unfortunate air of circularity. To eliminate it, far more will need to be said about how to determine whether some particular element of structure, used when applying a theory, is to be attributed to that theory's core or to some of its expansions.

Though I have only intuitions to offer on this subject, its importance may justify my briefly exploring them, beginning with a pair that Stegmüller and Sneed clearly share. Suppose that gravitational attraction varied as the inverse cube of distance or that the force of elasticity were a quadratic function of displacement. In those cases, the world would be different, but Newtonian mechanics would still be both Newtonian and mechanics. Hooke's law of elasticity and Newton's law of gravity therefore belong within the expansions of classical particle mechanics, not within the core which determines that theory's identity. Newton's second law of motion, on the other hand, must be located in the theory's core, for it plays an essential role in giving content to the particular concepts of mass and force without which no particle mechanics would be Newtonian. Somehow the second law is constitutive of the entire mechanical tradition which descends from Newton's work.

What, however, is to be said of Newton's third law, the equality of action and reaction? Sneed, followed by Stegmüller, places it in an expanded core, apparently because, from the late nineteenth century, it was irreconcilable with electrodynamic theories of the interactions between charged particles and fields. That reason, however, if I have identified it correctly, only illustrates what I previously referred to as an 'air of circularity'. The necessity of abandoning the third law was one of a number of recognized conflicts between mechanics and electromagnetic theory in the late nineteenth century. To some physicists, at least, the third law as well as the second thus seemed constitutive of classical mechanics. We may not conclude that they were wrong simply because relativistic and quantum mechanics had not yet been invented to take classical mechanics' place. If we did proceed in that way, insisting that the core of classical mechanics must contain all and only those elements common to all theories called Newtonian mechanics during the entire period that theory endured, then the equation of change-of-core with change-of-theory would be literally circular. The analyst who felt, as some physicists have, that special relativity was the culmination of classical mechanics, not its overthrow, might prove his case by definition alone, supplying, that is, a core restricted to elements common to both theories.

I conclude, in short, that before the Sneed formalism can be used effectively to identify and analyse episodes in which theory-change occurs by replacement, rather than simply by growth, some other techniques must be found to distinguish the elements in a core from those in its expansions. No problems of principle appear to block the way, for discussion of Sneed's formalism has already supplied important clues to their pursuit. What is needed, I take it, is an explicit and general articulation, within the formalism, of some widely shared intuitions, two of which were expressed above. Why is Newton's second law clearly constitutive of mechanics, his law of gravitation not? What underlies our conviction that relativistic mechanics differs conceptually from Newtonian in a way that, say, Lagrangian and Hamiltonian mechanics do not?[9]

In a letter responding to an earlier expression of these difficulties, Stegmüller has supplied some further clues. Perhaps, he suggests, a core must be rich enough to permit the evaluation of theoretical functions. Newton's second law is required for that purpose, he continues, but the third law and the law of gravity are not. That suggestion is precisely of the sort that is needed, for it begins to supply minimum conditions for the *adequacy* or *completeness* of a core. Even in so preliminary a form, furthermore, it is by no means trivial, for its systematic development may force the transfer of Newton's third law from the expansion of classical particle mechanics to its core. Though no expert in these matters, I see no way to distinguish inertial from gravitational mass (and thus mass from weight or force) without recourse to the third law. As to the distinction between classical and relativistic mechanics, remarks in Stegmüller's letter lead me to the following tentative formulation. Perhaps symbolically identical cores for the two theories could be found, but their identity would be only apparent. The two would, that is, make use of different theories of space-time for the specification of their non-theoretical functions. Obviously suggestions of this sort need work, but their ready accessibility is already reason to suspect that work will succeed.

4. REDUCTION AND REVOLUTIONS

Suppose now that techniques adequate to distinguish a core from its expansions were developed. What might it then be possible to say about the relation between changes of core and the episodes I have labelled scientific revolutions? Answers to that question will ultimately depend on the application of Sneed's reduction relation to theory-pairs in which one member at some time replaced the other as the accepted basis for research. No one to my knowledge has yet applied the new formalism to a pair of that sort,[10] but Sneed does tentatively suggest what such an application might try to show. Perhaps, he writes, the 'new theory must be such that the old theory reduces to (a special case of) the new theory' (Sneed, 1971, p. 305).

In his book, more clearly than in his contribution to this symposium, Stegmüller unequivocally endorses that relatively traditional suggestion, and he immediately employs it to eliminate what he calls the *Rationalitätslücken* in my viewpoint. For him as for many others these rationality gaps are found in my remarks on the incommensurability of pairs of theories separated by a revolution, in my consequent emphasis on the communication problems that confront adherents of the two, and in my insistence that those problems prevent any fully systematic, point-by-point comparison between them. (Stegmüller, 1973, pp. 14, 24, 165–169, 182f., 247–252.) Turning to these issues, I concede at once that, if a reduction relation could be used to show that a later theory resolved all problems solved by its predecessor and more besides, then nothing one might reasonably ask of a technique for comparing theories would be lacking. In fact, however, the Sneed formalism supplies no basis for Stegmüller's counter-revolutionary formulation. On the contrary, one of the formalism's main merits seems to me to be the specificity with which it can be made to localize the problem of incommensurability.

To show what is at issue, I begin by restating my position in a form somewhat more refined than the original. Most readers of my text have supposed that when I spoke of theories as incommensurable, I meant that they could not be compared. But 'incommensurability' is a term borrowed from mathematics, and it there has no such implication. The hypotenuse of an isosceles right triangle is incommensurable with its side, but the two can be compared to any required degree of precision. What is lacking is not comparability but a unit of length in terms of which both can be measured directly and exactly. In applying the term 'incommensurability' to theories, I had intended only to insist that there was no common language within which both could be fully expressed and which could therefore be used in a point-by-point comparison between them.[11]

Seen in this way, the problem of comparing theories becomes in part a problem of translation, and my attitude towards it may be briefly indicated by reference to the related position developed by Quine in *Word and Object* and in subsequent publications. Unlike Quine, I do not believe that reference in natural or in scientific languages is ultimately inscrutable, only that it is very difficult to discover and that one may never be absolutely certain one has succeeded. But identifying reference in a foreign language is not equivalent to producing a systematic translation manual for that language. Reference and translation are two problems, not one, and the two will not be resolved together. Translation always and necessarily involves imperfection and compromise; the best compromise for one purpose may not be the best for another; the able translator, moving through a single text, does not proceed fully systematically, but must repeatedly shift his choice of word and phrase, depending on which aspect of the original it seems most important to preserve. The translation of one theory into the language of another depends, I believe, upon compromises of the same sort, whence incommensurability. Comparing theories, however, demands only the identification of reference, a problem made more difficult, but not in principle impossible, by the intrinsic imperfections of translations.

Against this background, what I want first to suggest is that Stegmüller's use of the reduction relation is damagingly circular. Sneed's discussion of reduction depends upon an undiscussed early premise which I take to be equivalent to full translatibility. A necessary condition for the reduction of a theory T by a theory T' is a similar reducibility relation between the corresponding cores, K and K'. It in turn requires a reducibility relation between the partial potential models characterizing these cores. One requires, that is, a relation ρ which uniquely associates each member of the set M'_{pp} with a single member of the generally smaller set M_{pp}. Both Sneed and Stegmüller emphasize that the members of the two sets may be very differently described and that they may thus exhibit very different structures (Sneed, 1971, pp. 219f.; Stegmüller, 1973, pp. 145). Nevertheless, they take for granted the existence of a relation ρ sufficiently powerful to identify by its structure the member of M_{pp} which corresponds to a member of M'_{pp} with a different structure, described in different terms. That assumption is the one I take to be tantamount to unproblematic translation. Of course it eliminates the problems which, for me, cluster around incommensurability. But, in the present state of the literature, can the existence of any such relation simply be taken for granted?

In the case of qualitative theories, I think it clear that no relation of the sort ordinarily exists. Consider, for example, just one of the many counterexamples I have developed elsewhere (Kuhn, 1970, p. 107). The basic vocabulary of eighteenth-century chemistry was predominantly one of qualities, and the chemist's central problem was then to trace qualities through reactions. Bodies were identified as earthy, oily, metalline, and so forth. Phlogiston was a substance which, added to a variety of strikingly different earths, endowed them all with the luster, ductility, and so on common to the known metals. In the nineteenth century chemists largely abandoned such secondary qualities in favor of characteristics like combining proportions and combining weights. Knowing these for a given element or compound provided no clues to the qualities which had in the preceding century made it a distinct chemical species. That the metals had common properties could no longer be explained at all.[12] A sample identified as copper in the eighteenth century was still copper in the nineteenth, but the structure by which it had been modelled in the set M_{pp} was different from that which represented it in the set M'_{pp}, and there was no route from the latter to the former.

Nothing nearly so unequivocal can be said about the relation between successive theories of mathematical physics, the case to which Sneed and Stegmüller restrict their attention. Given a relativistic kinematic description of a moving rod, one can always compute the length and position functions that would be attributed to that rod in Newtonian physics.[13] It is, however, a special virtue of the Sneed formalism that it highlights the essential difference between that computation from relativity theory and the direct computation within Newtonian theory. In the latter case one starts with a Newtonian core and computes values directly, moving from application to application with the aid of specified constraints. In the former one starts with a relativistic core, and one moves through differently specified applications with the aid of constraints (on the length and time functions) that may also be different from the Newtonian. Only in a last step does one, by setting $(v/c)^2 \ll 1$, obtain numerical values that agree with the earlier computations.

Sneed underscores this difference in the penultimate paragraph of his book:

> the functions in the new theory appear in a different mathematical structure – they stand in different mathematical relations to each other; they admit of different possibilities of determining their values – than the corresponding functions in the old theory ... . Of course, it is an interesting fact that classical particle mechanics stands in a reduction relation to special relativity and that the mass functions in the theories correspond in this reduction relation. But this should not obscure the fact that these functions have different formal properties and, in this sense, they are associated with different concepts. (Sneed, 1971, pp. 305f.)

These remarks seem to me precisely right (compare Kuhn, 1970, pp. 100–102), and they suggest the following questions. Does not the reduction relation ρ between partial potential models demand an ability to relate the concepts, or formal properties, or mathematical structures underlying the M_{pp}''s and the M_{pp}'s prior to a computation of the concrete numerical values which those structures in part determine? Is it merely the fact that those computations can be done that has made the existence of the relation ρ between partial potential models seem so little problematic?

To this point I have dealt exclusively with the difficulties presented by the reducibility relation between cores. In the Sneed formalism, however, the specification of a theory demands specification not only of a core but also of a set of intended applications, I. The reduction of a theory T by a theory T' must therefore require some restrictions on the permissible relations between members of the sets I and I'. In particular, if T' is to solve all the problems resolved by T and more besides, then I' must contain I. In the general case of qualitative theories, it is doubtful that this containment relation can be satisfied. (Certainly, as the preceding remarks about chemistry indicate, T' does not always solve all the problems resolved by T.) But in the absence of even a crude formalization for such theories the issue is difficult to analyse, and I shall therefore restrict myself here to the intended applications of Newtonian and relativistic mechanics, a case in which intuitions, at least, are more highly developed. Its consideration will rapidly direct attention to what is, for me, the single most striking aspect of the Sneed formalism and also the one which most requires further development, not necessarily formal.

If relativistic mechanics is to reduce Newtonian mechanics, then the intended applications of the latter (i.e. the structures to which Newtonian theory is expected to apply) must be restricted to velocities small compared to the velocity of light. There is not, to my knowledge, a bit of evidence that any restriction of that sort entered the mind of a single physicist before the end of the nineteenth century. The velocities to be found in applications of Newtonian mechanics were restricted only *de facto*, by the nature of the phenomena that physicists actually studied. It follows that the historical class I, made up of intended, not simply of actual, applications, included situations in which velocity might be appreciable compared with that of light. To apply the reduction relations, those members of I must be barred, creating a new and smaller constructed set of intended applications which I shall label I_c.

For traditional formalisms this restriction on the intended applications of Newtonian mechanics is of no evident importance, and it has uniformly been disregarded. The reduced theory was constituted by the *equations* of Newtonian mechanics, and they remained the same, whether posited directly or derived from the relativistic equations in the limit. But in the Sneed formalism the reduced theory is the ordered pair $\langle K, I_c \rangle$, and it differs from the original $\langle K, I \rangle$ because I_c differs systematically from I. If the difference were only in the membership of the two sets, it might be unimportant, because the excluded applications would uniformly be false. A close look at the way in which membership in I_c and I is determined suggests, however, that something far more essential is involved.

Giving reasons for that evaluation requires a short digression concerning one last, especially striking parallel between my views and Sneed's. His book emphasizes that membership in the class of intended applications I cannot be given extensionally, by a list, because theoretical functions would then be eliminable, and theories could not grow by acquiring new applications. In addition, he expresses doubts that membership in I is governed by anything quite like a set of necessary and sufficient conditions. Asking how it is determined, he refers cryptically to the Wittgensteinian predicate 'is a game', and he suggests that basketball, baseball, poker, etc. "might be 'paradigm examples' of games". (Sneed, 1971, pp. 266–288, esp. p. 269.) Stegmüller's section, 'Was ist ein Paradigma?', considerably extends these points and calls explicitly upon similarity relations (*Ähnlichkeitsbeziehungen*) to explain how membership in I is determined. Many of you will know that learned similarity relations, acquired in the course of professional training, have also figured large in my own recent research. (Kuhn, 1970, pp. 187–191, 200f.; 1974. Note that neither Dr Sneed nor Professor Stegmüller had read these passages when their very similar views were developed.) I shall now very briefly extend and apply what I have previously said about them.

In my view, one of the things (perhaps sometimes the only thing) that changes in every scientific revolution is some part of the network of similarity relations that determines and simultaneously gives structure to the class of intended applications. Again, the very clearest examples invoke qualitative scientific theories. I have elsewhere, for example, pointed out that, before Dalton, solutions, alloys, and the compound atmosphere were usually taken to be *like*, say, metallic oxides or sulphates, and *unlike* such physical mixtures as sulphur and iron filings (Kuhn, 1970, pp. 130–135).[14] After Dalton, the pattern of similarities switched, so that solutions, alloys, and the atmosphere were transferred from the class of chemical to the class of physical applications (from chemical compounds to physical mixtures).

Lack of even a sketchy formalism for chemistry prevents my pursuing that example, but a change of much the same sort is visible in the transition from Newtonian to relativistic mechanics. In the former, neither the velocity of a moving body nor the velocity of light played any role in determinining the likeness between a candidate for membership in I and other previously accepted members of that set; in relativistic mechanics, on the other hand, both of these velocities enter into the similarity relation which determines membership in the different class I'. It is from the latter set, however, that the members of the constructed class I_c are selected, and it is that set, not the historical set I, that is used to specify the theory which can be reduced by relativistic mechanics. The important difference between them is not, therefore, that I includes members excluded from I_c, but that even the members common to the two sets are determined by quite different techniques and thus have different structures and correspond to different concepts. The structural or conceptual shift required to make the transition from Newtonian to relativistic mechanics is thus also required by the transition from the historical (and, in any usual sense, irreducible) theory $\langle K, I \rangle$ to the theory $\langle K, I_c \rangle$, constructed to satisfy Dr Sneed's reduction relation. If that result reintroduces a rationality gap, it may be our notion of rationality that is at fault.

These closing remarks should supply a fuller sense of the depth of my pleasure in Dr Sneed's formalism and in Professor Stegmüller's use of it. Even where we disagree, interaction results in significant clarification and extension of at least my own viewpoint. It is not, after all, a large step from Sneed's talk of 'different mathematical structure' or of 'different concepts' to my talk of 'seeing things differently' or of the gestalt switches that separate the two ways of seeing. Sneed's vocabulary gives promise of a precision and articulation impossible with mine, and I welcome the prospect it affords. But, with respect to the comparison of incompatible theories, it is entirely a prospect of things to come. Having insisted, in the first paragraph of this paper, that Sneed's new formalism makes important new territory accessible to analytic philosophy of science, I hope, in this closing section, to have indicated the part of that territory which most urgently requires exploration. Until it occurs, the Sneed formalism will have contributed little to the understanding of scientific revolutions, something I fully expect it will be able to do.

Princeton University and the
Institute for Advanced Study

NOTES

* First published in *Erkenntnis* **10** (1976), 179–199.

[1] The absence from traditional reconstructions of any step like that from a member of M_{pp} to its extension, the corresponding member of M_p, may help to explain my lack of success in persuading philosophers that normal science might be anything but a totally routine enterprise.

[2] Sneed and Stegmüller consider only theories of mathematical physics (only the mathematical parts of theories of mathematical physics, would be a better way to describe their subject). Therefore, they refer only to the role of constraints in the specification of theoretical *functions*. I add "or concepts" in anticipation of a needed generalization of the Sneed formalism. That Sneed himself believes that concepts are specified at least in part by mathematical structures which include constraints will appear below.

[3] A third example of the process (this time operating at the level of observation terms) which introduces language and empirical content in an inextricably mixed form is sketched in the last pages of my (1974). Its reappearance at all three traditional levels (observation terms, theoretical terms, and whole theories) seems to me of likely significance.

[4] Stegmüller, who rejects what he calls 'Sneed's Platonism', would put this point differently, and I find myself somewhat more comfortable with his approach. But its introduction here would call for additional symbolic apparatus irrelevant to this paper's main purposes.

[5] I could, for example, paraphrase a central theme of my forthcoming book on the history of the black-body problem in the following way. From 1900 through the publication of his *Wärmestrahlung* in 1906, the basic equations of mechanics and electromagnetic theory were in the core of Planck's black-body theory; the equation for the energy element, $\varepsilon = h\nu$, was part of its expansion. In 1908, however, the equation defining the energy element became part of a new core; equations selected *ad hoc* from mechanics and electromagnetic theory were in its expansion. Though there was sizable overlap between the equations included in the two *expanded* cores (whence much continuity), the structures of the theories determined by the two cores were radically distinct.

[6] Stegmüller suggests that my inability to resolve a cluster of difficulties presented by my position is due to my having accepted the traditional view of a theory as a set of statements (Stegmüller, 1973, pp. 14, 182). I shall express reservations about some of his illustrations of that suggestion below, but it is thoroughly relevant to the problem of continuity. Noticing that an equation or statement essential to a theory's success in a given application need not be a determinant of that theory's structure makes it possible to say much more about how new theories may be built from elements generated by their incompatible predecessors.

[7] A further sense in which one theory constrains another is indicated by the traditional view that the compatiblity of a new theory with others currently accepted is among the criteria legitimate to its evaluation.

[8] That a pan balance can be used to measure (inertial) mass can, of course, be justified only by resort to Newtonian theory. Presumably that is what Sneed has in mind when he argues (1971, p. 117) that mass must be theoretical because Newtonian theory may be

used to determine whether the design of a given balance is suitable for mass determination. That criterion (validation of a measuring instrument by a theory) is, I think, relevant to judgments of theoreticity, but it also illustrates the difficulties in making them categoric. Newtonian mechanics was, as a matter of course, used to check on the suitability of instruments for measuring time, and the ultimate result was the recognition of standards more precise than the diurnal rotation of the stars. I am not suggesting that Sneed's arguments for labelling time non-theoretical lack cogency. On the contrary, as already indicated, both they and their results fit my intuitions quite well. But I do think that the efforts to preserve a sharp distinction between theoretical and non-theoretical terms may by now be a dispensable aspect of a traditional mode of analysis.

My reservations about the full enforcability of Sneed's theoretical/non-theoretical distinction owe much to a conversation with my colleague C. G. Hempel. They were, however, initially stimulated by Stegmüller's repeated hints (1973, pp. 60, 231–243) that the distinction would require the construction of a strict hierarchy of theories. Terms and functions established by theory at one level would then be non-theoretical at the next higher one. Again, I find the intuition illuminating, but I see neither much likelihood of making it precise nor much reason for trying to do so.

[9] As the discussion to follow may indicate, the problem of distinguishing between a core and an extended core has a close counterpart in my own work: the problem of distinguishing between normal and revolutionary change. I have here and there used the term 'constitutive' in discussing that problem too, suggesting that what must be discarded during a revolutionary change is somehow a constitutive, rather than simply a contingent, part of the previous theory. The difficulty, then, is to find ways of unpacking the term 'constitutive'. My closest approach to a solution, still a mere *aperçu*, is the suggestion that constitutive elements are in some sense quasi-analytic, i.e. partially determined by the language in which nature is discussed rather than by nature *tout court* (Kuhn, 1970, pp. 183f.; 1974, p. 469n.).

[10] Sneed's examples are the reduction of rigid body mechanics by classical particle mechanics as well as the relations (more nearly equivalence than reduction) between the Newtonian, Lagrangian, and Hamiltonian formulations of particle mechanics. About them all, he has interesting things to say. But rigid body mechanics is, to a historical first approximation, younger than the theory by which it was reduced, and its conceptual structure is therefore straightforwardly related to that of the reducing theory. The relations between the three formulations of classical particle mechanics are more complex, but they coexisted without felt incompatibility. There is no apparent reason to suppose that the introduction of any but Newtonian mechanics constituted a revolution.

[11] When I first made use of the term 'incommensurability', I conceived the hypothetical neutral language as one in which any theory at all might be described. Since then I have recognized that comparison requires only a language neutral with respect to the two theories at issue, but I doubt that anything of even that more limited neutrality can be designed. Conversation discloses that it is on this central point that Stegmüller and I most clearly disagree. Consider, for example, the comparison of classical and relativistic mechanics. He supposes that as one descends through the hierarchy from classical (relativistic) particle mechanics, to the more general mechanics that lack Newton's second law, to particle kinematics, *and so on*, one will at last reach a level at which the non-theoretical terms are neutral with respect to classical and relativistic theory. I doubt the availability of any such level, find his "and so on" unilluminating, and therefore suppose that systematic theory comparison requires determination of the referents of

incommensurable terms.

[12] It would be wrong to dismiss this loss of explanatory power by suggesting that the success of the phlogiston theory was only an accident which reflected no characteristic of nature. The metals do have common characteristics, and these can now be explained in terms of the similar arrangements of their valence electrons. Their compounds have less in common because combination with other atoms leads to a great variety in the arrangements of the electrons loosely bound to the resulting molecules. If the phlogiston theory missed the structure of the modern explanation, it was primarily by supposing that a source of similarity was added to dissimilar ores to create metals rather than that sources of difference were subtracted from them.

[13] In Sneed's reconstruction, the field of particle kinematics is a low level theory which supplies the M_{pp}'s required to formalize all varieties of particle mechanics (the latter being determined by the various possible ways of adding force and mass functions to the M_{pp}'s). Classical particle mechanics emerges only with specialization to the subset M (of the M_p's) which satisfy Newton's second law. But that mode of division will not serve, I think, when Newtonian mechanics is to be compared with relativistic, for the two must be built up from different space-time systems and thus from different kinematics or differently structed M_{pp}'s. Lacking a developed formalism for special relativity, I shall therefore continue to treat a kinematics loosely, as part of the mechanics by which it is presupposed.

[14] Note that what I have here been calling a similarity relation depends not only on likeness to other members of the same class but also on difference from the members of other classes (compare Kuhn, 1974). Failure to notice that the similarity relation appropriate to determination of membership in natural families must be triadic rather than diadic has, I believe, created some unnecessary philosophical problems which I hope to discuss at a later date.

BIBLIOGRAPHY

Kuhn, T. S.: 1970, *The Structure of Scientific Revolutions*, 2nd. ed., Univ. of Chicago Press, Chicago, Ill.

Kuhn, T. S.: 1974, 'Second Thoughts on Paradigms,' in F. Suppe (ed.), *The Structure of Scientific Theories*, Univ. of Illinois Press, Urbana, Ill.

Sneed, J. D.: 1971, *The Logical Structure of Mathematical Physics*, D. Reidel Publishing Company, Dordrecht and Boston.

Stegmüller, W.: 1973, *Theorie und Erfahrung* (vol. II/2 of *Probleme und Resultate der Wissenschaftstheorie und Analytischen Philosophie*), Springer-Verlag, Berlin, Heidelberg, New York.

IV

PROGRAMME
OF THE 5th CONGRESS
(APPENDIX)

PROGRAMME OF THE
FIFTH INTERNATIONAL CONGRESS OF LOGIC,
METHODOLOGY AND PHILOSOPHY OF SCIENCE

London (Ontario) Canada, August 27 to September 2, 1975

WEDNESDAY MORNING, AUGUST 27

1000   Opening Ceremonies

WEDNESDAY AFTERNOON, AUGUST 27

Section I, Chairperson: S. C. KLEENE
1400–1500   Invited address
           Y. L. ERSHOV, Constructions 'By Finite'

Section VIII, Chairperson: D. L. HULL
           Contributed Papers
1400–1415   E. H. HUTTEN, Information and Meaning
1415–1430   N. P. REDDY, Information, Entropy and Knowledge
1430–1445   J. HUMPHRIES, Constructivity and Biological Adaptation
1445–1500   R. R. ROTH, The Individualistic Approach for Analysis of
           Hierarchic Levels in Biological Systems

Section XII, Chairperson: B. KEDROV
1400–1445   Invited address
           M. FREDE, The Origins of Traditional Grammar

Section IV, Chairperson: J. VAN HEIJENOORT
1515–1700   Symposium on the Concept of Set
           H. WANG, C. PARSONS

Butts and Hintikka (eds.), Historical and Philosophical Dimensions of Logic, Methodology and Philosophy of Science, 313–332.
Copyright © 1977 by D. Reidel Publishing Company, Dordrecht-Holland. All Rights Reserved.

Section XII, Chairperson: E. HIEBERT
 Invited addresses
1515–1600 W. R. SHEA, Galileo and the Justification of Experiments
1600–1645 J. MITTELSTRASS, Changing Concepts of the *A Priori*
1645–1730 M. BLEGVAD, Competing and Complementary Patterns of Explanation in Social Science
1730–1815 L. A. MARKOVA, Difficulties in the Historiography of Science

Section VI, Chairperson: I. LEVI
 Contributed Papers
1515–1530 S. SPIELMAN, Exchangeability and Objective Randomness
1530–1545 P. M. WILLIAMS, Coherence, Strict Coherence, and Zero Probabilities
1545–1600 P. BENIOFF, ZFC Models as Carriers for the Mathematics of Physics
1600–1615 R. CHUAQUI, A Model-Theoretical Definition of Probability
1615–1630 J. J. YOUNG, Frequencies, Long Runs and Unique Cases
1630–1645 R. SOLOMONOFF, The Adequacy of Complexity Models of Induction
1645–1700 K. SZANIAWSKI, On Sequential Choice between Hypotheses
1700–1715 E. W. CAMERON, Randomness and Mr Goodman's Paradox
1715–1730 R. ROSENKRANTZ, Inductive Generalization
1730–1745 B. L. LICHTENFELD, The Approach to a Logical Theory of Empirical Generalizations
1745–1800 N. STEMMER, Induction, Innate Capacities and Popper's Ideas
1800–1815 R. DACEY, J. KENNELLY, A Cognitivist Solution to Newcomb's Problem

Section IV, Chairperson: K. BERKA
 Contributed Papers
1700–1715 A. A. ZINOWYEW, Die Komplexe Logik und ihre Anwendungen

1715–1730   C. J. POSY, D. R. JOHNSON, F. W. SAWYER III, The Equilibrium Hypothesis in Mathematics

1730–1745   R. S. TRAGESSER, Platonism in Mathematics

1745–1800   D. VANDERVEKEN, A Formal Definition of the Class of the Logical Connectors of Pragmatics

1800–1815   P. IOAN, Linéaments pour une réhabilitation des principes de la pensée du point de vue formel

THURSDAY MORNING, AUGUST 28

Section V, Chairperson: I. NIINILUOTO

0900–1230   Symposium on Theory Change

        J. D. SNEED, Formal Inter-Theoretical Relations Within Empirical Theories of Science

        W. STEGMÜLLER, Accidental Theory Change and Theory Dislodgment: To What Extent Can Logic Contribute to a Better Understanding of the Dynamics of Theories

        T. S. KUHN, Theory Change as Structure Change: Comments on the Sneed-Stegmüller Approach

Section IV, Chairperson: E. AGAZZI

        Contributed Papers

0900–0915   S. SHAPIRO, On Church's Thesis

0915–0930   W. J. THOMAS, Church's 'Empirical Hypothesis?'

0930–0945   R. BOSLEY, Formulating Some Rules towards Unifying Two Concepts

0945–1000   C. F. KIELKOPF, A Case for the Limit Assumption

1000–1015   E. C. W. KRABBE, The Adequacy of Material Dialogue-Games

1015–1030   M. KANOVIČ, On One Problem of A. A. Markov

1030–1045   E. W. WETTE, On the Formalization of Productive Logic

1045–1100   Z. MIKELADZE, On a Sense of Quantifiers

1100–1115   R. I. WINNER, L. CHIARAVIGLIO, Programmatic Valuation Spaces

1115–1130   N. D. SIMCO, Logic and Ontology as Normative Sciences

1130–1145   C. YILDIRIM, The Logic of Criteria Formation

1145–1200   K. W. COLLIER, A. GASPER, What is the Place of Formal Logic in Philosophical Inquiry?

Mixed Section, Chairperson: A. I. UYEMOV
> Contributed Papers

0900–0915   F. VON KUTSCHERA, Indicative Conditionals–Section XI

0915–0930   D. SPASSOV, Logic Without Linguistics–Section XI

0930–0945   R. ABEL, Anthropocentric Aspects of Scientific Explanations–Section V

0945–1000   T. SETTLE, Moral Obligation as a Matter of Fact–Section V

1000–1015   A. PEICHENKIN, Some Ideas of the Philosophical Foundations of Quantum Chemistry–Section VII

1015–1030   J. WOODS, Metric 'Definite Descriptions'–Section IV

1030–1045   A. GIBBARD, W. HARPER, Conditionals and Decision with Act-Dependent States–Section VI

1045–1100   W. LEINFELLNER, A Marxian Foundation of Micro-Economics–Section X

1100–1115   N. BARRACLOUGH, Epistemological Axiomatization of Mathematics–Section IV

1115–1130   A. UYEMOV, Logical Apparatus of Systems Research–Section V

THURSDAY AFTERNOON, AUGUST 28

Section II, Chairperson: S. FEFERMAN
1400–1500   Invited address
> R. SOLOVAY, Large Cardinal Axioms and their Applications

Section X, Chairperson: S. NOWAK
1400–1500   Invited address
> V. J. KELLE, E. S. MARKARIAN, Types of Social Knowledge and the Problem of the Integration of Sciences

Section VIII, Chairperson: A. LINDENMAYER
1400–1445   Invited address
> G. STENT, Explicit and Implicit Semantic Content of Genetic Information

Section II, Chairperson: G. SACKS
  Contributed Papers
1515–1530   M. A. TAITSLIN, Existentially-Closed Structures
1530–1545   L. BUKOVSKY, On a Hypothesis of K. Namba Concerning the Closed Unbounded Ideal
1545–1600   A. G. DRAGALIN, Kripke's Scheme and Church's Thesis of Instuitionistic Analysis
1600–1615   W. C. POWELL, An Axiomatization of the Theory of Heyting-valued Models Generated by Order Topologies
1615–1630   Y. GAUTHIER, Intuitionnisme et théories mathématiques locales
1630–1645   B. A. KUŠNER, On the Extending Constructive Functions
1645–1700   J. T. SMITH, Group Theoretic Characterization of Metric Geometries of Arbitrary Dimension
1700–1715   ALEX S. YESHENIN-VOLPIN, Theory of Modalities and the Theory of Algorithms

Section VII, Chairperson: C. A. HOOKER
  Contributed Papers
1515–1530   P. A. BOWMAN, The Principle of Relativity and Conventionalism
1530–1545   P. MITTELSTEDT, Conventionality in Special Relativity
1545–1600   K. C. DELOKAROV, Gnoseological Bases of Theory of Relativity and Philosophy of Empiricism
1600–1615   J. B. MOLCHANOV, The Notions 'Now', 'Present', and 'Simultaneity' in Different Conceptions of Time
1615–1630   I. F. ASKIN, Sur la direction du temps
1630–1645   R. A. ARONOV, Discreteness of Space and Time and the Planck Relation
1645–1700   H. BARREAU, Les fondements rationnels de la notion de temps
1700–1715   P. KIRSCHENMANN, Are Symmetry Principles Devoid of Physical Content?
1715–1730   A. TURSUNOV, Concerning Foundations of Cosmology

Section VI, Chairperson: G. A. PEARCE
  Contributed Papers
1515–1530   J. SUTULA, The Impossibility of a Refutation of Fatalism
1530–1545   J. AGASSI, Determinism: Metaphysical versus Scientific

1545–1600   G. TRAUTTEUR, Operational Freewill for Deterministic Agents

1600–1615   G. LUCEY, C. LUCEY, On Causal Contribution

1615–1630   B. PYATNITSYN, On the Question of the Classification and Elimination of Indeterminacies

1630–1645   S. KLEINER, Meaning Variance in Moderation

1645–1700   K. KOSHEVOJ, Sciences Gnoseologically Interacting

Section XII, Chairperson: J. M. NICHOLAS
Contributed Papers

1715–1730   C. EISELE, The Role of Modern Geometry in the Philosophical Thought of C. S. Peirce

1730–1745   B. DYNIN, Has Philosophy of Science Still a Future?

1745–1800   W. A. WASSEF, Factors Affecting the Development of Science and Their Methodological Implications in Science Education

1800–1815   A. CORNELIS, Social Criticism and the Tragic Sense of Life: Structural Isomorphism in the Search for Rationality

FRIDAY MORNING, AUGUST 29

Section X, Chairperson: R. W. BINKLEY
0900–1100   Symposium on Justice and Social Choice
A. SEN, BENGT HANSSON

Section III, Chairperson: H. RASIOWA
Invited addresses

0900–0945   K. ČULIK, Basic Concepts of Computer Science

0945–1030   E. ENGELER, Structural Relations between Problems and Programs

1030–1115   A. SALWICKI, Algorithmic Logic, a Tool for the Investigation of Programs

Mixed Section, Chairperson: K. SZANIAWSKI
Contributed Papers

0900–0915   I. CARLSTROM, A Model Theory for the Quantifier 'Almost All'–Section VI

0915–0930   R. HILPINEN, On the Concept of Plausibility–Section VI

0930–0945 w. HEITSCH, Zur Rationalen Methodenkonstruktion in der Forschung–Section V

0945–1000 A. P. SHEPTULIN, Analysis and Synthesis in the Light of Knowledge Development–Section V

1000–1015 D. MUSTER, Les limites de la documentation scientifique contemporaine. En prospective–La priorité du fait–Section V

1015–1030 s. GALE, Arguments for a Question-Answering Theory of Scientific Inquiry–Section V

1030–1045 A. MENNE, Zur Theorie des Irrtums–Section V

1045–1100 K. BERKA, Materialistic Foundations of Measurement Theory–Section V

1100–1115 T. BALL, Rational Explanation Revisited–Section V

1115–1130 A. LYON, Laws of Nature–Section V

Mixed Section, Chairperson: I. HACKING
Contributed Papers

0900–0915 v. A. SMIRNOV, A New Form of the Deduction Theorem for P, E, and R–Section I

0915–0930 s. K. THOMASON, Modal Operators and Functional Completeness–Section I

0930–0945 v. KOSTJUK, Analytical Tableaux in Modal Calculi: A New Approach–Section I

0945–1000 E. NEMESSEGHY, A Model for Lukasiewicz Modal Logic–Section I

1000–1015 E. E. LEDNIKOV, Some Peculiarities of Hintikka's Semantics for Modalities–Section IV

1015–1030 J. WILLIAMSON, Time and Possible Worlds–Section IV

1030–1045 w. LENZEN, Probabilistic Interpretations of Epistemic Concepts–Section IV

1045–1100 E. K. VOJSHVILLO, An Experience of an Informal Analysis of the Logical System E (To the Problem of Explication of the Notion of Logical Entailment)–Section I

1100–1115 P. FOULKES, Strict Modality–Section IV

Section X, Chairperson: A. A. STARCHENKO
Contributed Papers

1100–1115 K. SEGERBERG, The Assignment Problem as a Problem of Social Choice

1115–1130   c. POPA, A Logical Theory of Cooperation
1130–1145   F. VANDAMME, Falsification in Economic Theory
1145–1200   M. PANT, H. BIST, Recent Attitudes to Science in Developing Countries

FRIDAY AFTERNOON, AUGUST 29

Section III, Chairperson: Y. V. MATIJASEVIČ
1400–1500   Invited address
            G. E. SACKS, RE Sets Higher up

Section X, Chairperson: S. PEPPER
            Contributed Papers
1400–1415   W. F. BARR, An Analysis of Typologies in Social Science
1415–1430   Z. A. JORDAN, A Logical Analysis of the Concept of Ideal Type in Social Science
1430–1445   D. PAPINEAU, Ideal Types and Empirical Theories in the Social Sciences
1445–1500   T. R. MACHAN, Defending Science in Social Science
1500–1515   E. S. MARKARIAN, On the Mechanism of Integration of Social and Natural Sciences

Section XI, Chairperson: H. SCHNELLE
            Invited addresses
1515–1600   D. WUNDERLICH, Some Problems in Speech Act Theory
1600–1645   A. KASHER, What are Pragmatical Representations: Foundations of Philosophical Pragmatics

Section V, Chairperson: A. BURKS
            Invited addresses
1515–1600   R. WOJCICKI, Towards a General Semantics of Empirical Theories

1600–1645   c. a. hooker, Methodology as a Component of Systematic Philosophy of Science
1645–1730   m. v. popovich, Identity by Sense in Empirical Sciences
1730–1815   e. agazzi, Subjectivity, Objectivity and Ontological Commitment in Empirical Sciences

Mixed Section, Chairperson: p. d. asquith
            Invited addresses
1515–1600   a. a. starchenko, Knowledge, Conviction and Belief: Conceptual Analysis–Section X
1600–1645   m. omelyanovsky (Read by V. V. Tselischehev) Axiomatics and the Search for Fundamental Principles and Concepts in Physics–Section VII
1645–1730   s. t. meljuchin, Philosophical Principles in the Theoretical Foundations of Physical Sciences–Section VII

Section I, Chairperson: w. craig
            Contributed Papers
1515–1530   j. p. seldin, Proof Normalizations and Generalizations of Glivenko's Theorem
1530–1545   m. n. bezhanishvili, On the Rejection of the Formulae of the First-Order Predicate Logic
1545–1600   a. oberschelp, Logic in Set-Theoretical Language
1600–1615   k. bing, Subformula Properties of the Natural Deduction Systems ND and NE
1615–1630   n. tennant, Natural Deduction for Restricted Quantification, Identity and Descriptions
1630–1645   r. a. pliuškevičius, On the Replaceability of the Restricted Induction Axiom
1645–1700   w. schwabhäuser, Non-Finitizability of a Weak Second-Order Theory
1700–1715   s. r. kogalivskii, Some Locality Criteria for Higher-Order Formulas
1715–1730   g. s. nadiu, An Abstract Calculus of Superior Order
1730–1745   a. mekler, Theories with Models of Prescribed Cardinalities

Section XI, Chairperson: H. HIŻ
         Contributed Papers
1645–1700   T. BALLMER, A Syntactic and Semantic Solution for the
            Bach-Peters Paradox
1700–1715   P. PETERSON, On the Logical Representation of Specific
            Noun Phrases
1715–1730   R. ZUBER, Noun Binding
1730–1745   R. M. MARTIN, On the Logic of Prepositions
1745–1800   J. D. ATLAS, A New Principle of Linguistic Semantics:
            Do Not Multiply Scope Ambiguities Beyond Necessity
1800–1815   KUNO LORENZ, Words and Sentences: A Pragmatic Ap-
            proach to the Introduction of Syntactic Categories

SATURDAY MORNING, AUGUST 30

Section XI, Chairperson: J. HINTIKKA
0900–1200   Symposium on Prospects for Transformation Grammar
            J. BRESNAN, Transformations and the Categories of
            Syntax
            S. PETERS, In Consequence of Speaking
            J.-R. VERGNAUD, Formal Properties of Phonological
            Rules

Section VII, Chairperson: N. CARTWRIGHT
         Invited addresses
0900–0945   I. A. ACKCHURIN, The Methodology of Physics and Top-
            ology
0945–1115   J. BUB, W. DEMOPOULOS, Symposium on the Interpreta-
            tion of Quantum Mechanics
1115–1200   J. EARMAN, Leibnizian Space-Times and Leibnizian
            Algebras

Section I, Chairperson: C. PARSONS
         Invited addresses
0900–0950   H. BARENDREGT, The Type-Free Lambda Calculus
0950–1040   A. S. TROELSTRA, Axioms for Intuitionistic Mathematics
            Incompatible with Classical Logic. (Read by J. Mos-
            chovakis)
1040–1130   M. PERETJAT'KIN, Constructive Models. (Read by Y. L.
            Ershov)

Section XII, Chairperson: R. LAUDAN
            Contributed Papers
0900–0915 V. M. DELGADO, The 'Consequencia' in the Works of Spanish Logicians 1470–1550
0915–0930 K. K. CHAKRABARTI, Some Non-Syllogistic Forms in Early Nyaya Logic
0930–0945 H. PETERS, Pre-Tarskian Distinctions
0945–1000 T. MESSENGER, Minimal Notations for PC
1000–1015 A. SCHLISSEL, An Interesting Relationship between Analytic Functions and Divergent Series
1015–1030 H. A. GEVORKIAN, On Alternative Characteristics of Scientific Knowledge
1030–1045 H. SEIGFRIED, On the Principle of Scientific Research
1045–1100 C. H. SEIGFRIED, Context and Experience
1100–1115 M. ODAGIRI, The Philosophical Foundation of the Present Revolution of Physical Sciences
1115–1130 E. M. BARTH, The Constants of Homeological Thought as Exemplified by Homeopathic Pharmacology

SATURDAY AFTERNOON, AUGUST 30

Section IV, Chairperson: Y. L. ERSHOV
1400–1500 Invited address
            A. A. MARKOV, On the Method of Staircase Semantical System in Constructive Mathematical Logic. (Read by Y. V. Matijasevič)

Section X, Chairperson: J. van EVRA
            Contributed Papers
1400–1420 E. A. MOUTSOPOULOS, Historiologie philosophique et philosophie de l'histoire
1420–1440 C. SAVARY, Reasons and Causes in History
1440–1500 K. ROBSON, Causality and Chronicles
1500–1515 M. KURNSEV, La science comme phénomène social

Section IX, Chairperson: H. A. SIMON
1515–1800 Symposium on the Status of Learning Theories

N. MACKINTOSH, Conditioning as the Perception of Causal Relationships

P. SUPPES, A Survey of Contemporary Learning Theories

A. EHRENFEUCHT and J. MYCIELSKI, Learnable Functions

Section VII, Chairperson: M. HESSE
1400–1500 Invited address
J. WHEELER, How Did the Universe Come into Being?

Section VII, Chairperson: T. di FRANCIA
Contributed Papers
1500–1515 E. MACKINNON, The Origin of Matrix Mechanics
1515–1530 E. W. STACHOW, Semantic Foundation of Quantum Logic
1530–1545 L. G. ANTIPENKO, Quantum Theoretical Object and Logical Theory of Types
1545–1600 B. Z. MOROZ, V. J. KREINOVICH, Comments on the Notion of State in Nonrelativistic Quantum Mechanics
1600–1615 G. M. HARDEGREE, Compatibility and Relative Compatibility in Quantum Mechanics
1615–1630 H. I. BROWN, Primary and Secondary Qualities in Quantum Mechanics
1630–1645 G. TORALDO di FRANCIA, M. L. DALLA CHIARA, On the Dividing Line between Deterministic and Indeterministic Physics
1645–1700 K. SHRADER-FRECHETTE, Conceptual Difficulties in High-Energy Physics
1700–1715 P. S. DYSHLEVY, On the Main Point of the Problem of Reality of Modern Physics
1715–1730 L. SOFONEA, Les hypostases du concept de 'zero' dans la pensee de la physique moderne

Section II, Chairperson: Y. GAUTHIER
Invited addresses and Contributed Papers
1515–1600 K. KUNEN, Closure Ordinal Spectra for Countable Theories–Invited address
1600–1645 L. W. SZCZERBA, Elementary Interpretability–Invited address
1645–1730 J. BAUMGARTNER, Ineffability Properties of Cardinals II–Invited address
1730–1745 L. W. MILLER, Characterization of Takeuti Ordinal Diagrams–Contributed Paper

1745–1800 D. CENZER, An Axiom which Implies Pre-Wellordering
(π_1^2)–Contributed Paper

1800–1815 R. WOLF, Formally Intuitionistic Set Theories with
Bounded Formulas Decidable–Contributed Paper

SUNDAY MORNING, AUGUST 31

Chairperson: P. SUPPES

0900–1200 Intersectional Symposium on Identifiability Problems
H. A. SIMON, Identifiability and the Status of Theoretical
Terms
V. N. SADOVSKY and V. A. SMIRNOV, Definability and
Identifiability: Certain Problems and Hypotheses
M. PRZEŁĘCKI, On Identifiability in Extended Domains
V. RANTALA, Prediction and Identifiability

Section VIII, Chairperson: B. van FRAASSEN
Invited addresses

0900–0945 I. T. FROLOV, Organic Determinism and Teleology in
Biological Knowledge

0945–1030 A. LINDENMAYER, Axiomatizing the Development of
Multicellular Organisms

1030–1115 D. HULL, The Ontological Status of Species as Evolutionary Units

Section III, Chairperson: E. ELCOCK
Contributed Papers

0900–0915 I. A. LAVROV, Computable Numberings

0915–0930 G. GERMANO, A. MAGGIOLO-SCHETTINI, Sequence to Sequence Partial Recursive Functions

0930–0945 S. H. STAHL, Primitive Recursive Ordinal Definability
and the Constructible Hierarchy

0945–1000 H. RASIOWA, Completeness Theorem for Extended Algorithmic Logic

1000–1015 O. COSMA, The Convergence of Intuitive-Algebraic
Method: The Finite Automata Modification Theorem

1015–1030 J. van LEEUWEN, An Algebraic Theory of Formal Languages

1030–1045 H. THIELE, Ein Erweiterter Aussagenkalkül und seine Anwendung auf die Semantik von Programmiersprachen

1045–1100 R. C. HOPSON, Cobol Programs for Generating Counter-examples

Section XII, Chairperson: F. L. HOLMES
Contributed Papers

0900–0915 L. LINDHOLM, The Wild Spirit of Realism

0915–0930 S. L. GOLDMAN, John Dee's Mathematical Preface

0930–0945 J. T. BLACKMORE, What Philosophy did Galileo Hold?

0945–1000 L. MOHLER, J. CORCORAN, Arab Geometers on Euclid's Concept of Ratio

1000–1015 R. D. PALMER, Some Epistemological Implications of 17th-century Science

1015–1030 A. F. CHALMERS, Copernicus and the Harmony of the Universe

1030–1045 U. STEINVORTH, Marx's Dialectical Method

1045–1100 J. J. GRADY, Watts and Faraday: On entertaining an Hypothesis

1100–1115 E. H. MADDEN, Chauncey Wright and Paradigm Shifts

1115–1130 K. J. DYKEMAN, The Principles of Induction: J. S. Mill and C. S. Peirce

1130–1145 J. BRADLEY, The Bridge from Sense-Perception to Science

1145–1200 N. G. MIKHAI, The Scientific Prerequisites of Neorationalism

1200–1215 J. THEAU, Le temps de la mécanique classique: sa structure et sa construction conceptuelles

SUNDAY AFTERNOON, AUGUST 31

Section XI, Chairperson: R. HILPINEN
Contributed Papers

1400–1415 H. PARRET, Illocutionary Force Representations

1415–1430 G. SOBLE, Searle on Reference and Atomism

1430–1445 G. P. MELNIKOV, Sources of Systematic Linguistics

1445–1500   H. H. LIEB, On Relating Theories of Language, Theories of Grammars and Grammars
1500–1515   V. T. STERBA, Explanation in Linguistics and Computer Science: A Case of Converging Evolution

Section X, Chairperson: M. BLEGVAD
1515–1600   S. NOWAK, Two Methodological Approaches to the Construction of Theories of Culture
1600–1800   J. HARSANYI, J. WATKINS, J. J. LEACH, Symposium on Rationality in Social Sciences

Section IV, Chairperson: L. J. COHEN
1515–1800   Symposium on the Philosophy of Logic
            D. KAPLAN, I. HACKING, V. TSELISHCHEV

Mixed Section, Chairperson: L. W. SZCERBA
            Invited addresses
1515–1600   Y. V. MATIJASEVIČ, Some Pure Mathematical Results Inspired by Mathematical Logic
1600–1645   I. A. LAVROV, Computable Numberings

MONDAY MORNING, SEPTEMBER 1

0900–1200   Intersectional Symposium on the Concept of Matter and its Development
            Chairperson: R. E. BUTTS
            B. KEDROV, On the Development of the Concept of Matter
            E. MCMULLIN, Matter and Activity in Newton

Section VI, Chairperson: M. V. POPOVICH
            Invited addresses
0900–0945   W. C. SALMON, Inductive Inference and Statistical Explanation

0945–1030 T. FINE, An Argument for Comparative Probability
1030–1115 A. TVERSKY, Causal Thinking in Judgment Under Uncertainty
1115–1200 I. NIINILUOTO, On the Truthlikeness of Generalizations

Section I, Chairperson: M. MORLEY
    Contributed Papers
0900–0915 P. HAJEK, Projective Classes of Models in Observational Predicate Calculi
0915–0930 K. A. BOWEN, On Intuitionistic Connectives
0930–0945 M. F. MARTINEZ, Sobre Correspondencias entre Fórmulas Cuantificadas y Productos de Fórmulas no Cuantificadas
0945–1000 C. A. INFANTOZZI, Sur les treillis de Skolem

MONDAY AFTERNOON, SEPTEMBER 1

Section IX, Chairperson: B. FARRELL
1400–1500 Invited address
    O. L. ZANGWILL, Consciousness and the Brain

Section XII, Chairperson: M. FREDE
    Contributed Papers
1400–1415 J. CORCORAN, Aristotle on Underlying Logics of Science
1415–1430 J. T. KEARNS, Aristotelian Quantifiers
1430–1445 A. PREUS, Geometrical Explanations in Aristotle 'Progression of Animals'
1445–1500 L. MOHLER, J. CORCORAN, An Ancient Generic Theory?

Section IX, Chairperson: B. FARRELL
    Invited addresses
1515–1600 V. P. ZINCHENKO, E. M. MIRSKY, Methodological Problems in the Psychological Analysis of Activity
1600–1645 R. TUOMELA, Causation of Action

Section II, Chairperson: P. MARTIN-LÖF
    Invited address
1515–1600 W. A. J. LUXEMBURG, Nonstandard Analysis and its Applications

1600–1745 Symposium on Category Theory
                S. FEFERMAN, Categorical Foundations and the Foundations of Category Theory

Section V, Chairperson: W. G. SALMON
                Contributed Papers
1515–1530 L. DARDEN, N. M. ROTH, The Unity of Science: Interfield Theories
1530–1545 L. S. GAGNON, What is a Scientific Field?
1545–1600 N. POLE, Why Value Conflicts do not Appear in Science
1600–1615 M. S. SLUTSKY, Determination of the Forms of Interconnection between Philosophy and Natural Science by the Level of their Development
1615–1630 E. LASHCHYK, Discoveries as Indices of Progress in Science
1630–1645 V. S. GOTT, The General Scientific Character and Development of Categories of Materialistic Dialectics
1645–1700 S. R. CARPENTER, On the Difference between Science and Technology
1700–1715 F. RAPP, Technological and Scientific Knowledge
1715–1730 R. MATTESSICH, Epistemological Consequences of Artificial Intelligence and Modern Systems Theory
1730–1745 V. HINSHAW JR., The Perceptual World and the Physical World

Section IX, Chairperson: M. PRZEŁĘCKI
                Contributed Papers
1700–1715 W. ROZEBOOM, The Enigma of Unboundedly Reactive Systems
1715–1730 R. CHAMPAGNE, A Weak Partial Ordering of the Noetic Field of Heuristic Intentionality Structures
1730–1745 C. MARE, La creativite et la creation scientifique dans l'epistemologie et la psychologie roumaine contemporaine
1745–1800 L. O. KATTSOFF, The One and the Many in Physics and Psychotherapy
1800–1815 H. BARENDREGT, Proposal of a Change in Attitude in Experimental Psychology

TUESDAY MORNING, SEPTEMBER 2

Section VI, Chairperson: R. GIERE
0900–1045   Symposium on the Concept of Randomness
C. P. SCHNORR, R. JEFFREY

Section V, Chairperson: R. S. COHEN
Contributed Papers
0900–0915   HERMANN LEY, Methodological and Substantial Problems Concerning a Multilevel Model of the Development of Science
0915–0930   M. A. FINOCCHIARO, Foundations of the Historiography of Science
0930–0945   P. MARCHI, Lakatos' Methodology and Metaphysical Bias
0945–1000   J. A. DOODY, Hanson and Radical Meaning Variance
1000–1015   W. B. JONES, A Note on the Comparability of Theories
1015–1030   S. G. HARDING, Three Problems for Popperian Rationality
1030–1045   W. BERKSON, Bartley's Theory of Rationality
1045–1100   J. N. HATTIANGADI, After Verisimilitude
1100–1115   L. BRISKMAN, Methodology and Metaphysics
1115–1130   A. BELSEY, Towards A Social Philosophy of Science?
1130–1145   N. LACHARITÉ, L'epistemologie et la socio-critique des sciences: leurs constructions respectives de la science comme objet
1145–1200   F. ROSSINI, R. I. WINNER, Explanation and Revolution
1200–1215   Y. A. MAMCHUR, On the Logical Reconstruction of Scientific Change

Section VIII, Chairperson: R. R. ROTH
Contributed Papers
0900–0915   C. WITTENBERBER, Polar Systems and Relations in Biology
0915–0930   S. RIGA, D. RIGA, The Methodology of Time Research in Biology
0930–0945   E. J. CROMBIE, Arbitrariness in Biological Classification
0945–1000   V. LOKAJÍČEK, Scaling and Measurement in Medicine
1000–1015   C. WHITBECK, The Concept of Syndrome in Medical Science

Section I, Chairperson: W. SCHWABHÄUSER
             Invited addresses
1515–1615 J. BARWISE, Some Eastern Two-Cardinal Theorems
1630–1730 P. MARTIN-LÖF, Meaning in Mathematics

Section V, Chairperson: W. ESSLER
             Contributed Papers
1515–1530 I. B. MIKHAILOVA, The Methodological Significance of
           the Analysis of Model Notions for the Solution of the
           Problem of Pictorialness in Modern Science
1530–1545 I. TUDOSECU, La finalité dans la perspective de la
           cybernétique
1545–1600 J. CASSIDY, J. GEORGE, Metaphors and Theoretical En-
           tities
1600–1615 J. STACHOVÁ, Models, Similarity, Metaphor
1615–1630 T. R. GIRILL, Micro-Parts
1630–1645 M. V. POPOVICH, Proof, Sense, and Indiscernibility in
           Empirical Sciences
1645–1700 J. GÖTSCHL, Science and Metatheoretical Criteria Sys-
           tems
1700–1715 P. FITZGERALD, Meaning in Scientific Theory
1715–1730 A. KAMLAH, An Improved Definition of 'Theoretical in a
           Given Theory'
1730–1745 S. PETROV, Concise Theoretical Terms
1745–1800 C. W. JOHNSON, The Swanson-Achinstein Triviality Ar-
           gument Regarding Formalists' Theoretical Modelling
1800–1815 P. KITCHER, Asymmetries in Explanation

Section VII, Chairperson: J. BUB
             Contributed Papers
1515–1530 H. R. POST, The Misuse of Models
1530–1545 J. I. KULAKOV, Theories der Physikalischen Strukturen
           und das 6. Problem
1545–1600 P. A. BOGAARD, Stability and Structure in the Concep-
           tual Foundations of Chemistry
1600–1615 G. KLIMASZEWSKY, Zur Leninistischen Funktion des
           Gedankenexperiments in der Physik
1615–1630 A. POLIKAROV, Interaction of Physics and other Sciences
           at the Methodological Level
1630–1645 R. O. KURBANOV, On the Methodological Importance of
           the Category 'Interaction' in Physical Cognition

INDEX OF NAMES*

*Names listed in the bibliographies are not included in this index.

THE UNIVERSITY OF WESTERN ONTARIO
SERIES IN PHILOSOPHY OF SCIENCE

A Series of Books on Philosophy of Science, Methodology, and Epistemology
published in connection with the University of Western Ontario Philosophy of
Science Programme

Managing Editor:

J. J. LEACH

Editorial Board:

1. J. Leach, R. Butts, and G. Pearce (eds.), *Science, Decision and Value.*
Proceedings of the Fifth University of Western Ontario Philosophy Collo-
quium, 1969. 1973, vii + 213 pp.
2. C. A. Hooker (ed.), *Contemporary Research in the Foundations and
Philosophy of Quantum Theory.* Proceedings of a Conference held at the
University of Western Ontario, London, Canada, 1973. xx + 385 pp.
3. J. Bub, *The Interpretation of Quantum Mechanics.* 1974, ix + 155 pp.
4. D. Hockney, W. Harper, and B. Freed (eds.), *Contemporary Research in
Philosophical Logic and Linguistic Semantics.* Proceedings of a Conference
held at the University of Western Ontario, London, Canada. 1975,
vii + 332 pp.
5. C. A. Hooker (ed.), *The Logico-Algebraic Approach to Quantum
Mechanics.* 1975, xv + 607 pp.
6. W. L. Harper and C. A. Hooker (eds.), *Foundations of Probability Theory,
Statistical Inference, and Statistical Theories of Science,* 3 Volumes. Vol. I:
*Foundations and Philosophy of Epistemic Applications of Probability
Theory.* 1976, xi + 308 pp. Vol. II: *Foundations and Philosophy of Statistical Inference.* 1976, xi + 455 pp. Vol. III: *Foundations and Philosophy of
Statistical Theories in the Physical Sciences.* 1976, xii + 241 pp.

8. J. M. Nicholas (ed.), *Images, Perception, and Knowledge*. Papers deriving from and related to the Philosophy of Science Workshop at Ontario, Canada, May 1974. 1977, ix+309 pp.
9. R. E. Butts and J. Hintikka (eds.), *Logic, Foundations of Mathematics, and Computability Theory*. Part One of the Proceedings of the Fifth International Congress of Logic, Methodology and Philosophy of Science, London, Ontario, Canada, 1975. 1977, x+406 pp.
10. R. E. Butts and J. Hintikka (eds.), *Foundational Problems in the Special Sciences*. Part Two of the Proceedings of the Fifth International Congress of Logic, Methodology and Philosophy of Science, London, Ontario, Canada, 1975. 1977, x+427 pp.
11. R. E. Butts and J. Hintikka (eds.), *Basic Problems in Methodology and Linguistics*. Part Three of the Proceedings of the Fifth International Congress of Logic, Methodology and Philosophy of Science, London, Ontario, Canada, 1975. 1977, x+321 pp.
12. R. E. Butts and J. Hintikka (eds.), *Historical and Philosophical Dimensions of Logic, Methodology and Philosophy of Science*. Part Four of the Proceedings of the Fifth International Congress of Logic, Methodology and Philosophy of Science, London, Ontario, Canada, 1975. 1977, x+336 pp.